EYES ON THE SKY

FRANCIS GRAHAM-SMITH

EYES ON THE SKY

A Spectrum of Telescopes

OXFORD
UNIVERSITY PRESS

OXFORD
UNIVERSITY PRESS

Great Clarendon Street, Oxford, OX2 6DP,
United Kingdom

Oxford University Press is a department of the University of Oxford.
It furthers the University's objective of excellence in research, scholarship,
and education by publishing worldwide. Oxford is a registered trade mark of
Oxford University Press in the UK and in certain other countries

© Francis Graham-Smith 2016

The moral rights of the author have been asserted

First Edition published in 2016

Impression: 1

Published in the United States of America by Oxford University Press
198 Madison Avenue, New York, NY 10016, United States of America

British Library Cataloguing in Publication Data
Data available

Library of Congress Control Number: Data available

ISBN 978-0-19-873427-7

Printed in Great Britain by
Clays Ltd, St Ives plc

PREFACE

In 1947, when I joined Martin Ryle in research on radio emission from the Sun, the only real astronomers were those who actually looked through telescopes. Optical astronomy, built on centuries of tradition and slow development, took little notice of the rest of the spectrum. The peak of emission from most stars lay in the visible spectrum, and there seemed little to be gained by exploring longer or shorter wavelengths. Larger telescopes like the 100-inch Hooker and the 200-inch Hale were already extending our view further into the Universe, and above all so much data could be recorded on a photographic plate that astronomers needed no other technology. Richard Woolley, who was appointed Astronomer Royal in 1956, was one of the first to appreciate the significance of radio emission from the Sun and the Milky Way, but nevertheless famously said that he wanted no electronics in his telescope domes; furthermore, he saw no point in space research. In October 1957 the first orbiting spacecraft, Sputnik I, was launched, and was tracked by radio astronomers. It seemed that this event changed everything overnight, but the revolution in astronomy which this book celebrates was already well under way. Three factors were at work: people, electronics, and digital computers.

Four centuries ago, Galileo first turned a telescope to look up at the night sky. His discoveries opened the cosmos, revealing the geometry and dynamics of the Solar System. The new astronomies, stretching over the whole spectrum from radio to gamma rays, have again transformed our understanding of the whole Universe. The transformation is evident in a sequence of popular books from eminent professors of astronomy in the University of Cambridge. Sir Robert Ball (1840–1913) wrote *Star-Land*— essentially his Royal Institute lectures of 1887, in which astronomy is about the geometry of the sky and a description of the Solar System. Sir

Arthur Eddington (1882–1944) wrote *The Nature of the Physical Universe* (1928)—a physicist's look at the new geometry of General Relativity, the internal structure of stars, quantum theory, and the thermodynamics of the expanding Universe; his contemporary Sir James Jeans (1877–1946) provided a mathematician's view of cosmology in *The Mysterious Universe* (1930); Stephen Hawking (b.1942) wrote *A Brief History of Time* (1988), which addresses the origin of the Universe; and Sir Martin Rees (b.1942, now Lord Rees), whose research had been on quasars and cosmology, wrote, among several popular books, a survey *From Here to Infinity* (2011). This sequence shows a dramatic progression in our understanding of the Universe. My aim in this book is to show how we built and used the telescopes that opened the new windows on the Universe.

There are now entire subjects within astronomy which were unimaginable before our new 'eyes on the sky' were opened: three examples are pulsars, galaxies, and the cosmos itself. All three depend on observations made across the whole electromagnetic spectrum, from radio through millimetre waves, infrared, visible light, ultraviolet, X-rays, to gamma rays. The techniques are very different, but there are unifying characteristics of telescopes designed for these various regimes. My intention in describing these apparently diverse telescopes is to emphasize their common characteristics, and how they must differ in adapting to the very wide range of wavelengths and photon energies in the spectrum. My own experience started in radio astronomy, expanding into optical astronomy through the design of telescopes for the optical observatory on La Palma in the Canary Islands, and brief excursions into cosmic rays and space research. But it was research on the Crab pulsar, which radiates pulses over the whole spectrum, that introduced me to the diverse realms of infrared, X-rays, and gamma rays. How do telescopes work in these unfamiliar territories? How has the astronomical community adapted to new ways of working, involving organizations of hundreds of people, remote observatories on mountaintops and on spacecraft, and massive computer-generated databases?

Bringing together the basics and the practicalities of observing over the whole electromagnetic spectrum has been a fascinating task. I have been greatly helped by colleagues at Jodrell Bank Observatory, and particularly by Prabu Thiagaraj, a visitor from the Raman Institute in Bangalore, who is helping design the Square Kilometre Array. I hope that our interest in telescope techniques and the people who put them to such good use will be widely shared.

FRANCIS GRAHAM-SMITH, 2015

CONTENTS

LIST OF FIGURES

Figures

LIST OF PLATES

Plates

1. The first reflecting telescope, built by Isaac Newton in 1668.
   © World History Archive/Alamy Stock Photo
2. Joseph von Fraunhofer demonstrating the spectroscope.
   Wikicommons
3. The four 8-metre telescopes of the Very Large Telescope (VLT) at Cerro Paranal, Chile.
   ESO/H.H.Heyer
4. The Mauna Kea observatories, Hawaii.
   Richard Wainscoat/Gemini Observatory/AURA/NSF
5. Images of Uranus obtained with the Keck telescopes on Mauna Kea.
   Lawrence Sromovsky, University of Wisconsin-Madison/W.W. Keck Observatory
6. The Sloan Digital Sky Survey telescope.
   Fermilab Visual Media Services www.sdss.org
7. The Hubble Space Telescope at its release from the Space Shuttle *Discovery* in 1990.
   NASA/Smithsonian Institution/Lockheed Corporation
8. The Hubble Ultra Deep Field.
   NASA/Smithsonian Institution/Lockheed Corporation
9. The James Webb Space Telescope.
   NASA/Northrop Grumman
10. The spectacular spiral galaxy M31 in Andromeda.
    NASA/JPL-Caltech/K. Gordon (University of Arizona)
11. The Eta Carina Nebula imaged by Herschel in three infrared wavelengths.
    ESA/PACS/SPIRE/Thomas Preibisch, Universitäts-Sternwarte München, Ludwig-Maximilians-Universität München, Germany

The publisher and the author apologize for any errors or omissions in the above list. If contacted they will be pleased to rectify these at the earliest opportunity.

1

GALILEO OPENS THE SKY

The First Telescope

Four hundred years ago Galileo published a book, *Siderius Nuncius* (*The Starry Messenger*), in which he announced the first astronomical discoveries made by using a telescope. He was a pioneer, an example to us all: he made his own telescope, and used it to observe the sky. What he saw was the first glimpse of the nature of the cosmos rather than merely the geometric pattern of the stars and planets. This was a double step forward; not only did he describe a real three-dimensional Universe, but he introduced the very idea of using an instrument to supplement the evidence of the unaided eye. It took some persuasion before his observations were accepted as anything more than a creation of his telescope itself.

Galileo could not possibly know that his telescope was using only a small part of the information reaching the Earth in a wide spectrum of radiation. Light is an electromagnetic wave, and we now use other parts of a very wide spectrum of these waves, from radio through X-rays to gamma rays, to observe the sky and describe the Universe. It is as if Galileo opened the first small window in a dark house, to be followed by generations of astronomers devising, building, and using telescopes and looking through new windows covering the whole of the spectrum. The new eyes with which we now see the Universe are the subject of this book.

For centuries before Galileo, astronomers were concerned with the geometry of the sky and the motions of heavenly bodies across it. They regarded the sky as a map drawn on the surface of a celestial sphere, with the Earth at the centre. Remarkably accurate observations of the positions

1

of the planets across this map were fitted into geometric models which accounted for their complex and unpredictable motion. Copernicus, in his 1543 book *De Revolutionibus Orbium Coelestium*, famously provided a simplification by proposing that the Sun, not the Earth, was the centre of the planetary system, but neither Copernicus nor any of his contemporaries could imagine investigating the actual nature of stars or planets. To the assiduous observers, using only their eyes and increasingly elaborate surveying instruments, they were points of light which moved according to strict rules.

Galileo (Figure 1) transformed astronomy, becoming the first astronomer to use a telescope rather than the unaided eye to observe the sky. As a professor of mathematics (at the University of Padua, near Venice), he was interested in optics, and set out to understand the theory of a new instrument known as the spyglass. This apparently had originated in the Netherlands, where in September 1608 Hans Lipperhey, a spectacle maker, had applied for a patent.[1] He was refused on the grounds that the idea was commonplace, and indeed there were spyglasses already on sale in that year in several European countries. These simple spyglasses used lenses that were used by spectacle makers to correct long and short sight, and in fact all that was needed to make a telescope was a pair of lenses that could be found in any spectacle-maker's workshop. The first lens would be convex—a converging lens as used for long sight—and the second would be concave—a diverging lens as used for short sight. These became known respectively as the objective and the eyepiece. The magnification was determined by the ratio of the strengths, or focal lengths, of these two lenses. The objective, which was the weaker lens, determined the length of the telescope tube, which was around 1 metre. The problem was that spectacle makers only made lenses with a limited range of strengths, so that most spyglasses had a magnification of only around x3.

Starting a tradition for astronomers by designing and making his own telescopes, Galileo realized that he needed a stronger concave eyepiece lens, and proceeded to make his own. He soon had a more powerful spyglass, with a magnification of x8. This was a dramatic improvement, with

Figure 1. Galileo presenting his telescope to the muses as the key to understanding the Solar System. An engraving in *Opere di Galileo Galilei*, published in Bologna in 1656.

obvious commercial and military uses; like many of his successors up to this day he realized that he should immediately establish his priority to the invention. He lost no time in demonstrating his new spyglass to the Doge and the Senate of Venice, showing that ships on the horizon became clearly visible, giving several hours' notice of their arrival in port. His reward was remarkably similar to the recognition given to modern academic astronomers: he was given tenure in his professorial appointment, and a rise in salary.

The next stage was an increase in magnification of the spyglass to x20, and the first observations of the Moon. Throughout a whole month, Galileo watched the Moon in all its phases, making sketches of its light and dark areas. His first great discovery came by looking carefully at the boundary, known as the 'terminator', between the sunlit and dark sides of the Moon, best seen at the first and third quarters of the lunar cycle. All through the early and late phases he saw detached areas of light and dark at this boundary, which he correctly interpreted as bright peaks and shadows of mountains and the edges of craters. Contrary to received wisdom at the time, which regarded everything in the sky as perfect and precisely geometrical, the surface of the Moon was not smooth but had mountains several miles high. This was powerful material, which started Galileo writing his famous *The Starry Messenger*. Even today, when we are inundated with new results and the new technologies that we will encounter in this book, this work ranks today alongside only a handful of outstanding papers in the entire history of astronomy.

The new telescope (as we will call the spyglass now that it was devoted to astronomy) must have been a well-constructed instrument. Modern hand-held binoculars, with a commonly used magnifying power of x8 or x10, are difficult to hold steady; anyone who has tried using a magnification of x20 or more knows that a steady mounting on a tripod is essential. Galileo says the same in his treatise, and he supplied rigid mountings with the dozens of telescopes that he eventually sold or gave to his patrons. Even under the best of conditions, images of stars were distorted and sometimes multiplied due to imperfections in the telescope lenses. Reducing the aperture at the objective ('stopping' in modern terminology) helped, and Galileo was soon able to pick out and count individual stars in asterisms such as the Pleiades. He found that the clouds in the Milky Way were made up of individual stars. The second part of *The Starry Messenger* sets out this discovery. The third part announced yet another discovery—the best of all—that revealed the true nature of the Solar System.

The Moons of Jupiter

The planet Jupiter has four conspicuous moons, which fortuitously can easily be seen with a low-powered telescope but which are just too faint to be seen by the naked eye. Galileo was the first to see them as star-like objects close to the planet. Only three were seen at first—two to the east and one to the west. He noticed that, improbably for stars, they were all located on a single almost straight line (Figure 2). A week later he looked again, and they were still in line—but they had moved! All three were now to the east of Jupiter. The fourth, which had been too close to the disc of the planet to be seen, emerged a few days later. They were all on the same straight line, or nearly so, but at different distances from Jupiter, moving at

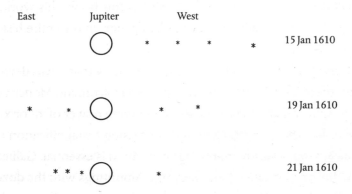

Figure 2. Jupiter's moons, as seen by Galileo on 15, 19, and 21 January 1610. These are three of the sixty-two sketches of the positions of the four moons as they orbit Jupiter.

different rates (Figure 2). Galileo followed all four moons throughout almost two months, making sixty-two sketches of their positions and showing that they were in orbit round Jupiter. He immediately realized that he had the key to understanding the whole Solar System. If Jupiter and Earth both have moons, then the Earth must be a planet, like Jupiter, and both must be planets. Instead of the Earth as the centre of the Solar System, the Sun must be the centre around which all the planets revolve, as proposed by Copernicus.

As an experimental physicist (although nominally a mathematician), Galileo easily grasped the significance of his observations. Others, however, were sceptical. Only the unaided eye could be trusted, and Jupiter's moons must therefore be produced by the telescope itself, an artefact with no fundamental meaning. Unfortunately not all of Galileo's telescopes worked well enough for the moons to be seen, or perhaps their mountings were not steady enough, and some independent observers could not (or would not) confirm his results. The news nevertheless spread rapidly, and proper publication became urgent; the result was *The Starry Messenger*, written, typeset, and printed all within two months of the first sighting of Jupiter's moons. The moons were first sighted on 7 January 1610, on 12 March a printed but unbound copy containing observations up to 2 March was sent to Cosimo II, Grand Duke of Tuscany, and unbound copies were available on 19 March.

Copernicus had already pointed out that if the planets are indeed orbiting the Sun, the planet Venus should exhibit phases like those of the Moon. This was surprisingly difficult to check, as the early telescopes produced distorted images, but Galileo demonstrated the effect in a series of observations in time for an announcement in December 1610, the same year as the publication of *The Starry Messenger*. There could now be no doubt that Copernicus was right, and also that the telescope was telling a true story.

The speed of publication of *The Starry Messenger* did not detract from quality, and this little book is as accessible and fresh today as ever.[2] It is the first astronomical paper which is centred round observations with a new scientific instrument rather than following the traditional geometrical or philosophical descriptions of the heavens. Again grasping the importance of his discoveries, Galileo made another move which resonates with modern scientists: he dedicated the four moons of Jupiter to the Medici family in Florence, his native city. He was already well known to Cosimo, the head of the family, who happened to have three brothers. The moons, one for each brother, became known as the Medicean Stars, and Galileo was assured the support of this most powerful family.

The Next Moves: Kepler

The design of the simple spyglass that inspired Galileo's telescope can still be found in some opera houses and theatres as an 'opera glass'. Notoriously, in this form it has a low magnification of only x2 or x3, whereas the binoculars in common use today for birdwatching and the like use a much improved system (Figure 3), introduced in 1611 by Kepler (he who sorted out the geometry of planetary orbits). This system, originally known as the Keplerian telescope, uses a convex lens rather than a concave lens as eyepiece. A larger magnification is easier to obtain with this system, with a larger field of view. A disadvantage is that the image is inverted, which

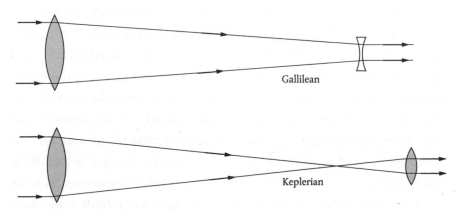

Figure 3. The arrangement of lenses in Galilean and Keplerian telescopes.

hardly matters for astronomy but must be overcome for terrestrial use by inserting an inverting prism between the two lenses.

All but the most elementary refracting telescopes must overcome the problem of chromatic aberration. Glass refracts and focuses light of different colours slightly differently, and without correction any image of a star would be surrounded by an out-of-focus coloured halo. The solution is to combine two different types of glass into a compound lens, cancelling out the aberration over a wide range of colours. Some massive glass lenses using this principle were eventually made for astronomical telescopes, the largest having a diameter of 40 inches. This was installed at Yerkes Observatory, founded in 1897 by George Ellery Hale (1868–1938). As we will see later, this was the end of an era for refracting telescopes, as Hale soon transferred his interest to reflecting telescopes, which can be made very much larger and which dominate the scene today.

Greater magnification was often required by users of these early telescopes. Since magnification is the ratio of the powers of the eyepiece and objective, a very low-powered objective was needed, necessitating an objective with a long focal length. This was taken to an extreme by observers such as Johannes Hevelius (1611–1687) of Danzig and the Dutch scientist Christiaan Huygens (1629–1695), who made some rather impractical telescopes up to 165 feet long (Figure 4). Despite the difficulties, these long telescopes were used successfully to observe the planets and enabled the discovery of several of their moons. Dominico Cassini (1625–1712), the first Director of the Paris Observatory, used successively longer telescopes (17 feet, 34 feet, 100 feet, and 136 feet) and discovered four satellites of Saturn. The Cassini spacecraft, at present (2015) still in orbit round Saturn, is very appropriately named after him.

Although the magnification of these early telescopes was already sufficient to bring previously unknown stars into view, it soon became evident that a larger aperture—that is, a larger objective lens—was needed if fainter objects were to be observed. This is quite simply because a larger aperture collects more light—a consideration that applies to all telescopes of whatever character. Larger and larger lenses were impractical. The solution was

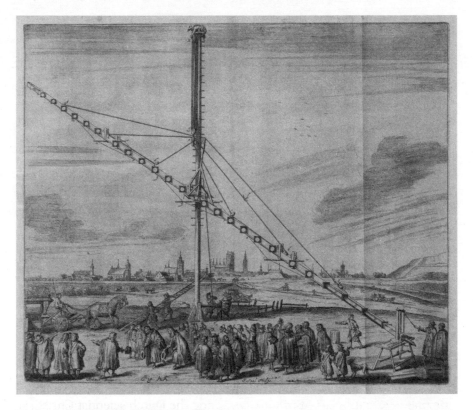

Figure 4. The 150-foot tubeless telescope constructed by the German astronomer Johannes Hevelius (1611–1687). The lenses were mounted in a wooden trough supported by a vertical 90-foot pole, and the instrument was operated by a team of workers with horses, ropes, and pulleys. (*Machina Coelestis*, 1673.)

the reflecting telescope, which we now see with ever-increasing size in modern astronomy.

Using a concave mirror instead of a lens as the objective of a telescope has the great advantage that all colours are reflected in exactly the same way, completely avoiding chromatic aberration. The geometry of the telescope changed radically, to solve the problem of arranging an eyepiece so that the telescope can be looked through. This led to new designs, which began to look like the giants of modern astronomy. Figure 5 shows the basic ideas.

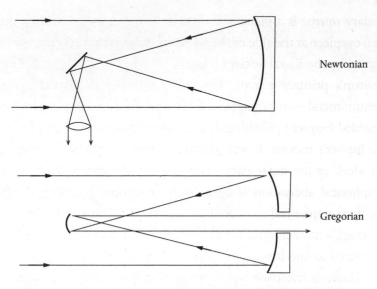

Figure 5. The Gregorian and Newtonian reflecting telescopes.

Gregory, Newton, and Herschel

The first design, published in 1668 by James Gregory, a Scottish mathematician, is now known as the Gregorian. In this design there is a focal point inside the telescope tube, as in Kepler's design but followed by a second, smaller mirror reflecting light back through a hole in the main mirror. Thus an eyepiece can be mounted at the end of the tube, allowing a direct view of the sky. The image is upright, so that the Gregorian is useful as a terrestrial telescope; for example, in surveying. Gregory was unable to construct such a telescope himself, and the first Gregorian was built by Robert Hooke in 1673. A similar design was published by Laurent Cassegrain in 1672, and many modern astronomical telescopes use this system, placing their main instrumentation at a focus usually called the Cassegrain, behind the main mirror.

The second design, by Isaac Newton and constructed by him in 1668, was the first reflecting telescope to be built. In this Newtonian system the secondary mirror is a plane mirror at an angle of 45°, deflecting the light into an eyepiece at the side of the telescope tube. Newton's first telescope is preserved by the Royal Society (Plate 1).

Newton's primary mirror was only 1.3 inches in diameter. Made of speculum metal—an amalgam of tin and copper—it tarnished rapidly and needed frequent polishing. It was, however, good enough for Newton to see Jupiter's moons. It was ground to have a spherical surface, which is not ideal, as the outer parts focus slightly differently from the centre. This spherical aberration is completely overcome by using a parabolic rather than a spherical surface. This was achieved in 1721 by John Hadley (1682–1744), a mathematician and a Fellow of the Royal Society, who is also remembered as the inventor of the navigation instrument known as the octant. Hadley's telescope was a Newtonian of 6 inches aperture. He also made Gregorian telescopes.

William Herschel (1738–1822) stands head and shoulders above other telescope makers of his time. He was born in Germany, and moved to England when he was 19 years of age. He and his family were very musical; he became director of the Bath Orchestra and composed many symphonies and works for soloists. Starting in Bath, where his house is now a museum, he built a series of reflecting telescopes with the clear purpose of systematically exploring the sky, concentrating on unusual objects such as double stars and nebulae. His best-known achievement was the discovery of the planet Uranus. Perhaps following the example of Galileo's naming of the four Medicean moons of Jupiter, he dedicated the new planet to King George III and named it the Georgium Sidus (Georgian Star). Again following Galileo, he sold many telescopes throughout Europe, and there are examples in most scientific museums. The Science Museum in London has an example of his largest mirrors, which reached 50 inches in diameter. They were made of speculum metal with some arsenic added to the tin and copper, which improved the reflecting surface. The largest mirror was used in his famous 40-foot telescope (Figure 6), constructed in the back garden of his house in Slough; a diagram of this telescope is part of the distinctive logo of the Royal Astronomical Society. His favourite instrument, with

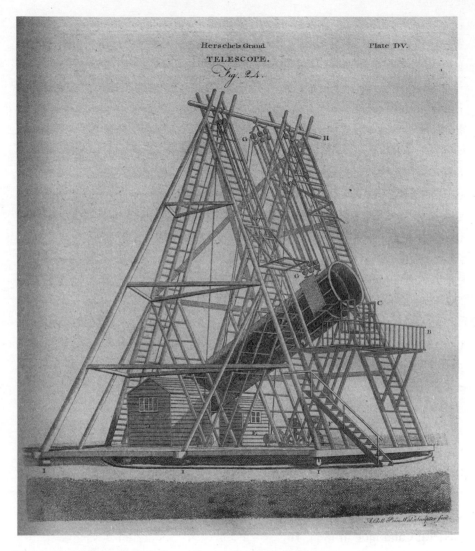

Figure 6. William Herschel's 40-foot reflecting telescope.

which he and his sister Caroline (1750–1848) carried out most of their survey work, used a smaller mirror, 6¼ inches in diameter.

The requirement for convenient and steady mounting of telescopes for serious and continuing surveys led Herschel to the rigid wooden structures which dominate the pictures of the 40-foot telescope. Most of his telescopes were moveable in elevation only; observations were made at a fixed elevation, watching the sky as it drifted across the field of view and making sketches and notes on objects as they appeared. Some, like the 40-foot, were mounted on a turntable—a mounting allowing movement in azimuth as well as elevation (or altitude). This mounting is now known as an altazimuth (alt-az, or az-el), and is commonly used in modern large optical and radio telescopes. Herschel's earliest telescopes followed the Newtonian pattern, but he later adopted the new idea of an offset focus, slightly tilting the primary mirror and moving the eyepiece to the rim of the telescope tube, so avoiding obstruction of the incoming light. Again, there is a modern example in the GBT radio telescope (see Chapter 8).

The case made today for any new telescope, ground-based or space-based, from radio to gamma ray, inevitably sets out a solid programme of work and adds a 'serependipity' clause, saying that the unknown and unpredictable discoveries may be the most valuable. Herschel's programme followed this pattern. He set out to survey the sky, cataloguing nebulae, but his most famous discovery was Uranus, followed by two moons for each of Uranus and Saturn, and several comets discovered by his sister Caroline. But most astronomers remember him for his determined and persistent survey of the whole of the northern sky, without the help of photography, compiling a catalogue of over 2,000 nebulae. This was extended by his son John Herschel (1792–1871), who observed the southern sky from near Cape Town, and who published the combined General Catalogue in 1864. This was the basis of the familiar New General Catalogue (NGC), compiled in 1880 by John Dreyer and still in use. Furthermore, William Herschel was the first to deduce from the proper motion of stars that the Solar System was moving against the stellar background, and he also suggested that the Milky Way was in the form of a disc of stars seen from its edge.

William Herschel set the standard in visual observing for future generations, but the mid-nineteenth century saw the introduction of spectroscopy and photography, on which modern astronomy depends.

The New Windows

Until the middle of the twentieth century, visible light was almost the only part of the electromagnetic spectrum available to astronomy. The visible spectrum—violet, indigo, blue, green, yellow, orange, red—seemed to be bounded at both ends by the atmosphere, which absorbed ultraviolet light and much of the infrared. Sunlight, and apparently most starlight, fitted nicely into the visible spectrum, and it seemed that our eyes, supplemented by photography, could gather most of the information that we would ever know about the Universe.

Apart from minor extensions into ultraviolet and the near-infrared, radio was the first new window to be opened. The atmosphere is no barrier to radio waves, even when it is cloudy; it provides ground-based astronomy with the only clear view of the sky, apart from visible light. But the way radio is used looks very different from conventional optical astronomy. The difference is that the wavelengths of radio are millions of times greater than those of light (Figure 7). Extending the spectrum in the other direction, to shorter wavelengths, takes us into the X-ray and gamma-ray regions. Here the wavelengths are shorter than those of light by a factor of thousands for X-rays, and by further factors of millions for gamma rays. The unaided eye is useless; new detector systems are essential, and telescopes look less and less like Galileo's or Newton's. Furthermore, opening

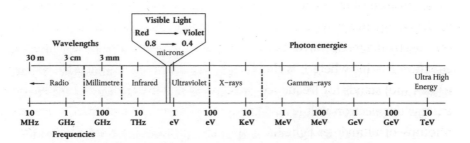

Figure 7. The electromagnetic spectrum, showing wavelengths and frequencies for the long wavelengths, and photon energies for the short wavelengths.

windows for X-rays and gamma rays needs spacecraft to take our telescopes above the atmosphere. Astronomy now looks even more unfamiliar. It is the aim of this book to show how the techniques of observing over the whole of the spectrum are, in practice, closely related.

Any part of the spectrum is characterized either by wavelength or frequency; the two are simply related, as their product is the velocity of all electromagnetic waves in free space, which is close to 300 million metres per second. Very small and very large numbers are unavoidable in covering the whole wide spectrum, but it is useful to think of the visible spectrum as a midpoint (Figure 7). At the long-wavelength radio end we usually think in terms of wavelengths, ranging from around 1 metre to 1 mm. In the other direction, beyond visible light, the wavelengths become so small that they can only be compared with the size of an atom or even an atomic nucleus. Descriptions in terms of frequency meet a similar difficulty. At the lowest-frequency end, radio, the frequencies are reasonably familiar, since they are used in broadcasting and communications. Here we are accustomed to descriptions in terms of units of megahertz (MHz), a million cycles per second, and the highest radio frequencies in terms of gigahertz (GHz), a factor of a thousand larger. But at the other end of the spectrum the numbers become so large that they are impossible to comprehend, and we use different descriptions with more meaningful numbers.

For the visible spectrum, and for X-rays, we can avoid the embarrassment of descriptions in terms of frequency beyond the mega- and giga-range. Instead of frequency we use a different approach in which we think of the energy in a wave arriving as discrete photons rather than a continuous electromagnetic wave. Using Planck's constant h in the fundamental equation $E = h\nu$, where E is the energy of a single photon and the Greek letter ν (nu) stands for frequency, we can use a convenient unit for photon energy: the electron volt (eV). The numbers now become easier, since the photons of ultraviolet light have an energy of a few electron volts. Proceeding up the range of energies, we reach X-rays, with photon energies measured in kilovolts (keV), and gamma rays at megavolts (MeV) and above. With this enormous range, it is surprising that we can find links between the astronomy of the opposite ends of the spectrum. As we will see, there are some objects in our complex Universe which make themselves known to us by radiating over the whole of the spectrum. First, however, we will follow the development of astronomy in the visible part of the electromagnetic spectrum, bearing in mind that this is a tiny fraction of the whole.

2

THE BIG REFLECTING
TELESCOPES

The modern era of large optical telescopes dates from the three great Californian telescopes: the 60-inch Hale and 100-inch Hooker at Mount Wilson Observatory, and the 200-inch Hale on Mount Palomar. This progression from 60 inch to 100 inch and to 200 inch was inspired by George Ellery Hale, whom we met in Chapter 1 as the founder of the Yerkes Observatory with its 40-inch refractor. The glass lens of the Yerkes refractor weighs half a tonne, and is still the largest primary telescope lens. The new large optical telescopes were reflectors, using concave mirrors and breaking a long tradition of using lenses started by Galileo.

It had been known since the pioneering efforts of Gregory, Newton, and Herschel that reflectors have a fundamental advantage over refractors: reflection is ideally independent of colour, avoiding the chromatic aberration that occurs in refraction at a lens surface. Their telescope mirrors were made of speculum metal. But attempts to make larger mirrors with the difficult technique of casting and polishing speculum metal were only partly successful. The largest reflecting telescope built since Herschel's time was the Leviathan of Parsonstown (Figure 8), built in 1845 by William Parsons, 3rd Earl of Rosse, at Birr Castle in Ireland. As with all its predecessors, its 72-inch mirror was made of speculum metal. From the late 1850s, speculum-metal mirrors began to be replaced by silvered glass mirrors, which became very popular, though it took quite some time for this method to be adopted for much larger instruments. Hale, however, in partnership with the outstanding telescope builder George Ritchey, broke

Fig. 8. Lord Rosse's Leviathan at Birr Castle. The telescope tube is mounted between masonry walls, and can be moved in elevation and through several degrees either side. The speculum-metal mirror weighs three tons.

through the barrier of technique and tradition with an entirely new approach using silvered glass mirrors for very large telescopes.

Most of Hale's scientific work was on the Sun, which he investigated using the new techniques of photography and spectroscopy. He regarded the Sun as a star which happened to be close enough to be investigated in detail, and he was driven by the conviction that the same methods could be applied to other stars if more powerful telescopes could be constructed. This required larger apertures, beyond the possibilities of glass lenses. The new generation of telescopes had to have mirrors which would be accurate enough to concentrate all the light entering a large aperture onto a photographic plate or the entrance slit of a spectrograph. Furthermore, photography and spectroscopy demand long exposure times, so the telescope had to maintain focus and alignment for several hours. Astronomy was entering the modern age in which telescopes for all its regimes, from radio wavelengths to gamma rays, depend on large apertures, accurate tracking over long exposures, sensitive recording, and spectroscopy.

Photography

Until the invention of the emulsion photographic plate, observing involved looking directly through an eyepiece. The objective lens or mirror was producing an image of the sky, and this was examined by eye, looking through another, smaller lens. Using a telescope as a camera is a simplification: basically only one lens or mirror is needed, with the photographic plate, or more recently an array of electronic detectors, such as the charge coupled device (CCD), replacing the eyepiece. The advantages are obvious; there is no need to measure or draw the image at the telescope, but overwhelmingly more importantly, the photograph could be far more sensitive through building up a faint image by using long observing times.

For practical purposes, photography can be considered to have begun with the introduction of Louis Daguerre's process in 1839. John Herschel was involved in the early development of photography, and it was probably he who invented the name. In 1839 he published a method of fixing images which was adopted by Henry Fox Talbot in his pioneering artistic photography. The new technology was soon taken up by astronomers. In 1840, John W. Draper obtained a photograph of the Moon; in 1845, Leon Foucault and Louis Fizeau obtained a photograph of the Sun; in 1850, a photograph of Vega was obtained by William Bond and John A. Whipple at Harvard College Observatory; during the 1850s, Warren De La Rue pioneered lunar photography; and the first photographic spectrum of a star, Vega, was obtained by Henry Draper in 1872. When David Gill, at the Cape Observatory in South Africa, photographed the Great Comet of 1882, he realized that he had also photographed many stars, and therefore began a programme of stellar photography. At the same time, Andrew Common in England and Henry Draper in America obtained photographs of stars and nebulae. Galileo had shown that the Milky Way is composed of myriads of stars, so were the small discrete clouds, which we now know are extragalactic nebulae, also star clouds, possibly looking smaller because they are further away? Common and Draper both began by photographing the Orion Nebula, and were widely recognized for their work: Common received the Gold Medal of the Royal Astronomical Society in 1884, and Henry Draper's name has been preserved as the inspiration for a widely used catalogue of stars, the HD Catalogue.

The advantages of photography were now obvious, and telescopes were adapted or newly built to accommodate the new technique. The transformation rapidly led to a major international effort to photograph and chart the whole sky, which involved using no fewer than eighteen similar telescopes in observatories around the world. But coordination on this scale proved to be difficult. After an enthusiastic start in 1887, the project matured notoriously slowly, and the resulting catalogue, known as the Carte du Ciel (Map of the Sky), was eventually published in 1962.

With a photographic plate replacing the human eye, the telescope became a camera. There was no need to build longer telescopes to obtain high magnification suitable for the human eye; fine angular detail could be recorded on a small linear scale on the photographic plate. Telescope apertures became larger, and are still growing larger more than a century later, but telescopes are not growing in length in the same proportion. Photographers will be familiar with the term focal ratio, or f-number—the ratio of focal length to lens diameter in their cameras. Telescopes are now commonly built with f-numbers as small as 3, instead of 10 or more in the prephotographic era. The new generation of 'Extremely Large Telescopes', which we describe later in this chapter, takes this trend even further, with f-numbers as low as 0.7.

Photographic plates and films can record images in microscopic detail, and the design of the telescope has to be matched to the scale of detail which can be recorded on the plate, or to the pixel size of the individual elements in modern detector arrays. The scale of the image at the focus of a camera depends only on its focal length, a longer focal length producing images which may be larger than necessary for the best sensitivity. A lens with a short focal length, producing an image concentrated on a single pixel, is often called a fast lens, since the concentrated image needs the shortest exposure. The human eye (Figure 9) is a good example of a camera in which the size of the sensitive cells in the retina is well matched to the smallest detail which can be distinguished. The focal length of the lens is 25 mm, and when the pupil is fully open the lens has a low focal ratio of 2, making the eye a 'fast' camera.

An overwhelming advantage of photography is that long exposure times can be used, accumulating light in a single image for as long as the telescope can be guided continuously on a selected object or area of the sky. A famous example was when Walter Baade (1893–1960) and Rudolf Minkowski (1895–1976) used the 100-inch Hooker telescope to observe very faint stars in an extragalactic nebula. Taking advantage of the dark sky during the blackout of World War II, they observed a small field of stars in the Andromeda Nebula continuously for two whole nights, and were able to pick out the Cepheid variables which they used to establish the distance

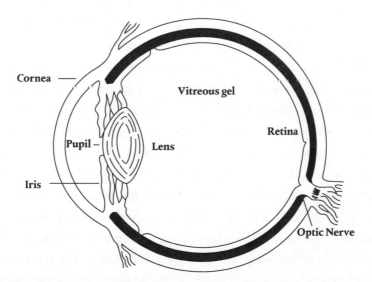

Fig. 9. The human eye. The focal length of the combined cornea and lens is 25 mm. A point source of light focuses to an image 0.02 mm across, matching the size of the light-sensitive cells on the retina. The size of the pupil, which in telescope terms is the aperture, is controlled by the iris.

scale of the Universe. An even longer exposure was achieved with the Hubble Space Telescope, using a CCD detector system—the modern replacement of the photographic plate. It observed continuously for eleven days to produce the deepest survey of the faintest and most distant galaxies.

Splitting the Spectrum

When in 1666 Isaac Newton used a prism to split a beam of sunlight into a coloured spectrum, he was exploring the nature of light rather than investigating its origin. It was not until 1802 that William Wollaston (1766–1828) investigated the spectrum of light from the Sun in greater detail. He noticed that there were dark lines in the solar spectrum; but despite the fact that he was a chemist and is credited with the discovery of two new elements (palladium and rhodium) he did not associate these lines with any chemical elements. At this time, Thomas Young (1773–1829) showed that light was behaving as a wave, like ripples on the surface of water, and that different colours corresponded to different wavelengths. Josef von Fraunhofer (1787–1826) used a spectroscope (Plate 2) to distinguish and list no fewer than 570 of these dark lines, assigning a wavelength to each, but again without associating them with known chemical elements. This important step was taken around 1859 by Gustav Kirchoff (1824–1887) and Robert Bunsen (1811–1899). They found that the wavelengths of bright lines in the laboratory spectra of heated elements such as iron and sodium exactly matched the wavelengths of the dark Fraunhofer lines.

An inspired amateur, William Huggins (1824–1910), now heard of the identification of solar spectral lines with known elements, and set out to use the new technique on stars and nebulae. He had an 8-inch refractor, for which he built a two-prism spectroscope (Figure 10), allowing the spread-out spectrum to be viewed directly (later, the spectroscope became a spectrograph, which recorded spectra photographically).

The most urgent question at this time was the nature of nebulae such as the familiar Orion Nebula. Were these bright clouds tightly packed collections

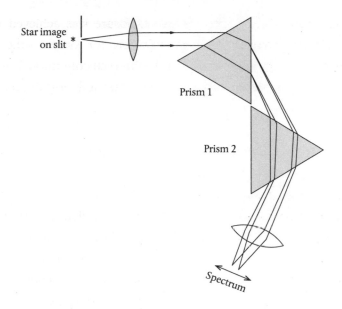

Fig. 10. The two-prism spectrograph used by William Huggins in 1862 in his discovery of spectral lines in starlight.

of stars like the Sun? The image of a nebula was placed on a slit at the spectroscope entrance, so that any spectral feature appeared as a line rather than a point (this is the origin of the term spectral line, which is used universally for all such features in any part of the spectrum, even in radio). Huggins had already in 1863 found dark lines in the spectra of fifty stars, the same pattern as the Fraunhofer lines in the Sun, so proving that stars are like the Sun and that the same chemical elements were present everywhere. Some of the nebulae proved to be totally different; the first of these was a planetary nebula—a symmetrical cloud surrounding an isolated star. The spectrum of this nebula contained several individual bright lines instead of a continuous spectrum with dark lines. This indicated that the nebula was a glowing cloud of gas, like the hot gases which showed spectral lines in the laboratory. Later work showed that the gas in these planetary nebulae is largely hydrogen and oxygen, lit up by ultraviolet light from a hot central star. Spectroscopy also eventually showed that the faint light from most of the faint isolated nebulae was starlight. These became known as extragalactic nebulae.

Splitting starlight into a spectrum is usually achieved today by using a diffraction grating. The basic idea is to pass a light beam through a grid of narrow parallel slits, closely spaced like the grooves on a CD or DVD recording disc. It is a common experience that light reflected from the surface of a CD is split into rainbow colours; the colour reflected in any direction depends on the wavelength in relation to the spacing of the grooves in the disc. Several very remarkable men were involved in the early-nineteenth-century history of the diffraction grating. Thomas Young (1773–1829)—a professional physician who is famous for his contribution to the decipherment of the Rosetta Stone—and Augustin-Jean Fresnel (1788–1827)—inventor of the Fresnel lens used in lighthouses—both worked out the relation between wavelength and diffraction angle. In 1785 David Rittenhouse—a self-educated craftsman who eventually became the first director of the United States Mint—was the first to construct a grating, using fifty very fine hairs stretched across a frame. Larger gratings, with thousands of slits or lines, were needed for high-resolution spectroscopy; these were made by precise ruling machines which cut lines into the surface of a silvered mirror. A ruled mirror grating could be used in two ways, either in reflection like the CD, or transmission through the clear slits in the mirror surface. Various more subtle forms are now used, including glass transmission gratings which are completely transparent but whose thickness is modified in a similar pattern of lines. However, these different types of diffraction grating all work in the same basic way: they all act by imposing a corrugation on a plane wave, with the result that a wave splits off at an angle to the incident light. This wave, like the light beam leaving a prism, is at a different angle for different coloured light, so forming a spectrum. This can then be scanned across a single detector, as in Figure 11, or focused onto a photographic plate or an array of detectors such as a CCD.

A diffraction grating may waste light by producing more than one diffracted beam. Efficient gratings can now be made, and easily reproduced, which concentrate a beam of light into a single spectrum. This beam is at an angle to the incoming light, which may be inconvenient. The solution is to

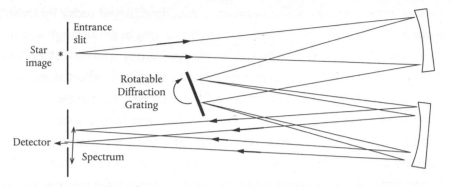

Fig. 11. A diffraction-grating spectrometer. This arrangement uses no lenses, and can therefore be used for ultraviolet light.

combine a grating and a prism, by forming the grating on one face of the prism (the combination has the happy name of grism). A thin grism may be inserted in front of the focal plane of a telescope, and all the images of stars in the field of view are then spread into tiny spectra.

Mounting the Monsters

The simplest way of mounting a telescope is on a single east–west axis, so that it can swing to any elevation from horizon to horizon. A survey of the whole sky is then achieved by setting the telescope at a fixed elevation for a whole night, and letting a strip of the sky drift by. This was used by William, Caroline, and John Herschel in their painstaking mapping of the whole sky, resetting the elevation at the start of every night. A telescope mounted in this simple way is known as a transit telescope. This became the basis of accurate sky surveys and time-keeping at the Royal Observatory, Greenwich.[1] The Airy Transit Circle, which was set up in 1850 by Sir George Biddell Airy,[2] defined the north–south meridian line at longitude zero. Visitors to the observatory stand astride this line with one foot in the eastern hemisphere and the other in the western hemisphere. This transit telescope was used to time the transit of the Sun, Moon, planets, and stars, watching their passage across a fine line in the focal plane and recording the time as shown by an accurate pendulum clock. The Cooke Reversible Transit replaced the Airy Transit Circle at Greenwich and was in use from 1936 to 1953, when the observatory moved to Herstmonceux Castle. In the nine-teenth century, smaller transit circles were used in many cities for determin-ing local time before the telegraph linked all clocks to Greenwich time. The only requirement for mounting these transit telescopes was a single bearing aligned precisely on an east–west line. But for a telescope to be capable of pointing to any region of sky at any time, and to follow the same spot through a long photographic exposure, is a more demanding problem.

Hale's three great reflecting telescopes are recognizably the predecessors of today's giant optical telescopes, apart from a striking difference in the way the telescopes were mounted and moved around the sky. Pointing a telescope to a star anywhere in the sky, and following a star as the Earth rotates, requires motion about two axes. This is done in two ways: the equatorial mounting and the altazimuth mounting. If the telescope is attached to a structure that can rotate on an axis which is aligned parallel to the rotation axis of the Earth, then a slow rotation of that support structure can compensate for the Earth's rotation. Movement about a single axis will then keep a star in the centre of the field of view. Stars at different distances from the North Pole can be followed by setting the telescope at a different angle on the rotating mount. This system of mounting the tele-scope on a polar axis is known as a polar or equatorial mount. An early example of a large reflecting telescope on a polar mount was the 48-inch reflector built in 1877 at the Paris Observatory (Figure 12), but with the eyepiece near the front of the massive tube (the Newtonian arrangement) it must have been difficult to use.

The polar mount was used in all three of Hale's telescopes. Designing a polar mount for a large telescope is a massive engineering challenge.[3] The mirror of the 200-inch telescope alone weighs 14.5 tons, and with its support system it must be held steady to better than 1 arcsecond as it follows a star across the sky. The telescope itself is slung inside the polar axis, which is a massive structure in the form of a horseshoe. The outside of the horseshoe runs on a very smooth oil-film bearing. Figure 13 is Russell

Fig. 12. The great 48-inch reflector built in 1877 at the Paris Observatory.

Porter's beautiful engineering drawing of the telescope in its dome. The huge horseshoe bearing dominates the picture.

The alternative mounting for large optical telescopes is the altazimuth mount, familiar in the mounting of large radio telescopes. The telescope moves on two axes; the elevation axis is like the bearing of a transit telescope, but the whole system is on a rotating horizontal platform

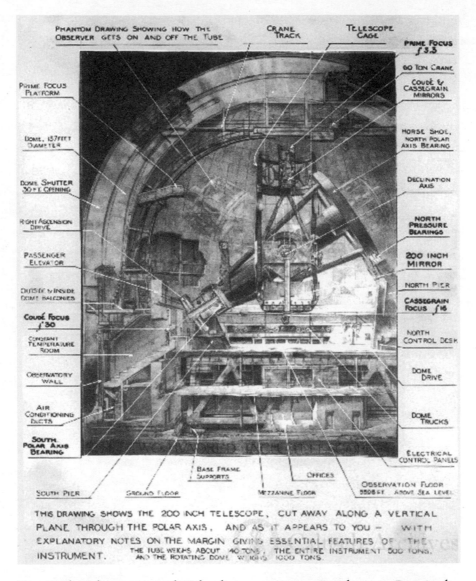

Fig. 13. The Palomar 200-inch Hale telescope on its equatorial mount. Drawing by Russell W. Porter, California Institute of Technology.

which moves on an azimuth bearing or on a circular rail track. The engineering of an altazimuth mount for a large telescope is much simpler than for a polar mount, but it is no longer possible to follow the movement of a star by a steady rotation on a single axis. The necessary movements on both axes are, however, easily computed and applied automatically to the separate drive motors.

An early example of a telescope on an altazimuth mounting was constructed by James Nasmyth, a pioneer of the machine tool industry who is known for his invention of the steam hammer. On retiring in 1856 at the age of 48 he designed and built a 20-inch reflector (Figure 14a) with a new arrangement for an observer (Figure 14b). In all previous designs the observer had to look through an eyepiece which moved inconveniently as the telescope tracked across the sky, sometimes precariously perched on a scaffold high above the ground. Nasmyth's design uses a second, smaller, mirror mounted at the entrance aperture, as in Gregory's innovation, but adding a third mirror deflecting the light beam through a right angle, so

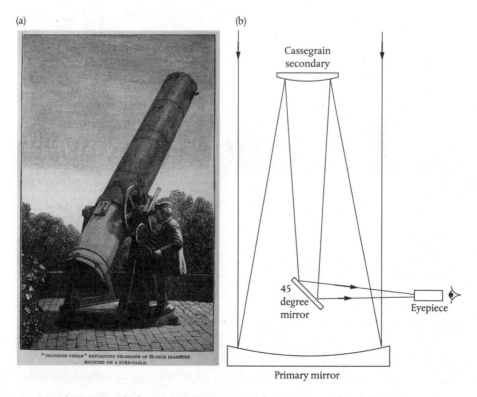

(a)

(b)

Cassegrain secondary

45 degree mirror

Eyepiece

Primary mirror

"TRUNNION VISION" REFLECTING TELESCOPE OF 20-INCH DIAMETER MOUNTED ON A TURN-TABLE.

Fig. 14. The Nasmyth focus. a) James Nasmyth sitting comfortably on a rotating platform. b) A 45-degree mirror sends the light along the elevation axis.

that it can be viewed via a hole through the telescope elevation bearing. The observer sits comfortably on the rotating platform, and moves round with the telescope as it rotates on the azimuth axis. The same system is adopted in modern telescopes, where the observer's position is replaced by a platform on which apparatus such as spectrometers can be mounted.

Covering the Sky: the Palomar Schmidt

The 100-inch Hooker and the 200-inch Hale telescopes dominated observational astronomy for half a century, opening up research into extragalactic nebulae and laying the foundations of modern cosmology. They were unrivalled not only for the light grasp of their huge apertures, but also for the quality of their stellar images and their ability to follow a target object for the long observing periods needed for spectrometry. They were particularly good at identifying objects such as radio sources, which turned out to be very distant galaxies and quasars, many of which proved to be at cosmologically significant distances. This type of telescope, however, could only cover a very small field of view at any one time. The way astronomy was developing, and continues to develop today, still requires the most sensitive observations of individual objects, but also requires a simultaneous survey of many objects within a large field of view. A telescope was needed which had a large enough aperture to detect faint and distant objects, while covering a field of view measured in degrees rather than minutes of arc. This was achieved with the Schmidt telescope.

Up to the time of the 100-inch Hooker, the cross-section of the mirror in all large telescopes had always been a parabola rather than a section of a circle. A spherical surface does not focus properly; rays reflected from the outer parts come to a focus closer to the mirror than the inner parts. This is called spherical aberration. The parabolic surface focuses perfectly, but only for a star on or close to the telescope axis. Telescopes with a secondary mirror, like the Gregorian, can improve on this by using a slightly different shape, a hyperboloid, for both the primary and the secondary mirrors.

This idea, introduced by George Ritchey and Henri Chrétien, provides good images over a wider field of view than the single paraboloid. Hale, however, was not convinced of the merits of the Ritchey–Chrétien design, and persisted in designing the 100-inch Hooker telescope with a parabolic primary mirror. Schmidt's design was a more radical solution to the problem, which was to revert to a simple spherical surface and deal with the spherical aberration with a separate thin lens of unusual profile.

Bernard Schmidt was a gifted optician, born in Estonia in 1879, who became an expert in grinding and polishing telescope lenses and mirrors, despite having lost his right hand in an accident with a home-made bomb. After he became associated with Hamburg Observatory in 1926, he met Walter Baade, who is now remembered as one of the greatest of observational astronomers. Baade pointed out the need for a camera telescope covering a large field, with a large aperture and with sharp images. Schmidt arrived at the solution in 1930. His scheme is shown in Figure 15.

Schmidt's solution was to use the symmetry of the spherical surface, and correct the focus by adding a thin lens at the centre of curvature of the spherical mirror. Unlike the lens of a refracting telescope, this corrector is almost a plane sheet of glass, but its thickness varies as shown exaggerated

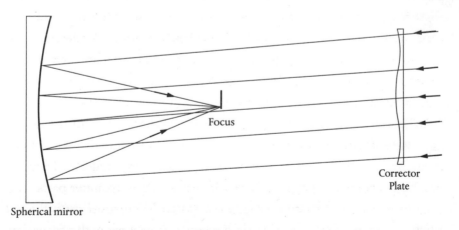

Focus

Corrector
Plate

Spherical mirror

Fig. 15. The Schmidt camera. The primary mirror has a spherical profile; the thin corrector plate located at the centre of curvature corrects the spherical aberration and brings a wide field of view into focus.

in the diagram. This changes the relative focal length for inner and outer rays, producing a perfect focus. The inspiration was to place this 'Schmidt plate' at the centre of curvature of the mirror, so that it acted very nearly in the same way for rays both on axis and at a large angle. As can be seen in Figure 15, this doubles the overall length of the telescope. Consequently, there are no very large Schmidt telescopes, as they would need large and expensive domes to house them. Another practical problem is that the image at the prime focus is on a curved plane, so the photographic plate has to be curved to match.

A 48-inch Schmidt telescope (now called the Oschin Schmidt) was built in the early 1950s at Mount Palomar Observatory, and a close copy was built at the Anglo-Australian Observatory (AAO).[4] These were used in collaboration to cover the sky in a complete survey, each 14-inch photographic plate covering a 6-degree square on the sky. Print copies of this monumental survey were distributed to observatories around the world, and the plates were scanned to produce a digitized version in 1994. With the survey complete, these two telescopes are now used for entirely different observations; the Oschin Schmidt is now equipped with a large array of CCD detectors used in repeated surveys, searching for transient phenomena, and the AAO Schmidt specializes in simultaneous spectroscopy of many objects spread over its large field of view. It is now engaged on a comprehensive compilation of spectra of stars in the Milky Way, providing redshifts which will be used to investigate the dynamics of the galaxy on large and small scales.

Spectroscopy at the AAO Schmidt uses an ingenious system devised by John Dawe and Fred Watson and originally installed at the prime focus of the Anglo-Australian Telescope. The idea was to measure the spectra of many galaxies simultaneously with the objective of finding their redshifts and hence their distances. Measuring these one at a time would be impossibly tedious and time-consuming, but a system was devised to feed light from up to a hundred objects simultaneously into a single spectrograph, while keeping their spectra separate. This is achieved by fibre optics. An array of fibres connects the images at the focus to a long slit at the spectrograph, so that the output of the spectrograph is a stack of spectra, with one line for each target object. Locating and fixing so many fibres in the focal plane is a complicated task, involving measuring the positions of the required objects and automating the placing of the fibres, which are held in place by small magnets on a metal plate.

The AAO Schmidt was equipped with this new multi-object spectrograph in 1982. The results were spectacular. The redshifts of more than 100,000 galaxies were measured, covering the whole of the southern sky. Figure 16 shows the magnitude of the task, showing all surveyed galaxies in the southern sky. The same technique was adopted for a redshift survey of galaxies in the northern hemisphere in the Sloan Digital Sky Survey (SDSS), which is described in the next chapter. The significance of the surveys is

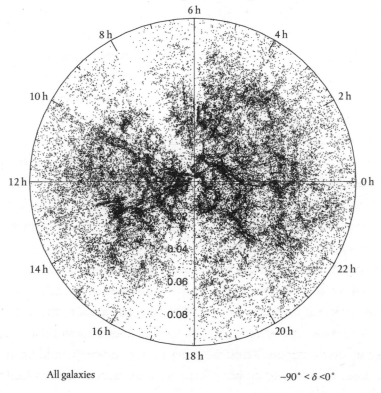

All galaxies $-90° < \delta < 0°$

Fig. 16. The local Universe. This map, covering the southern hemisphere, shows all galaxies out to redshift z = 0.1. The redshifts of all these galaxies were measured using the fibre-optic spectrograph of the Anglo-Australian Observatory Schmidt telescope,

hinted at by the structure which can be seen in the figure. Although this shows only a two-dimensional map, galaxies are obviously distributed in clumps and extended groups which can be related to the original structure of our Universe.

The AAO Schmidt and SSDS surveys showed an unexpectedly complete clustering of galaxies into clouds, streaks, and sheets. Relating these structures to the pattern of fluctuations in the early Universe, as discovered in the Cosmic Microwave Background, is a major task for cosmology, which we will discuss in Chapter 11.

The idea of simultaneous observations of many objects spread over a large field of view, introduced by the Schmidt camera, has had a major impact on the next generations of telescopes, as will be seen in Chapter 3.

Bigger and Better

The example of the Californian telescopes stimulated a revolution in observational astronomy. The possibilities of making major advances in almost any branch of astronomy, but especially extending our view of the cosmos beyond a local group of galaxies, were seen to be limited by the resources available to most observatories; furthermore, new departures in observing very faint objects demanded clear dark skies which were seldom encountered at the traditional home observatories. Expeditions to search for sites for the most demanding observing conditions focused on mountaintop sites in both hemispheres, leading to a concentration of the world's largest optical telescopes in Hawaii and the Canary Islands in the north, and Chile in the south. These new observatory sites are remote from the home observatories of most of their designers and users, so that funding and operating them are inevitably (and advantageously) undertaken by the communal efforts of a number of astronomers and their funding organizations, often from many different countries.

The second half of the twentieth century saw fundamental improvements in telescope design, with a new generation of reflectors—several of them with apertures of around 4 metres, and soon exploring new methods taking apertures up to 8 or 10 metres. Remarkably, the 200-inch Palomar telescope remained the largest for most of this time, the limitations on larger designs being due to the sheer weight of the main mirror and the consequent difficult engineering of the equatorial mounting. New concepts were needed to allow mirrors to be made with larger diameters. The last of the style that had become conventional were two 4-metre telescopes—one at the US National Observatory on Kitt Peak, followed by an improved version, the Anglo-Australian Telescope (AAT), at Siding Spring Observatory in Australia. The AAT, completed in 1974, has been remarkably productive, for two reasons. It was the first modern telescope in the southern hemisphere, and it was superbly engineered, with efficient optics and excellent instrumentation.

The 16-tonne main mirror of the AAT was cast at Owens-Illinois in the USA and figured at Grubb Parsons in Newcastle upon Tyne, UK. The material is Cervit—a low-expansion glass-ceramic composite. A secondary mirror at the prime focus returns light through a 1-metre diameter hole in the centre to the main focus behind the main mirror. This is the Cassegrain focus, where most of the instrumentation is mounted. The thickness of the main mirror is 63 cm. In modern terms this is a thick mirror, which relies on its own stiffness to maintain its shape at all angles. Despite its stiffness, the support for the mirror must be carefully distributed over its back and edges. Once a year or thereabouts the mirror, supported in its cell, must be removed from the telescope and placed in a vacuum chamber, where it is given a new reflecting coat of aluminium. So many factors conspire to make it almost impossible to build larger telescopes on this pattern that it is all the more surprising that the 200-inch Palomar telescope was built so early in this story and has been so successful.

The AAT has proved to be a very versatile telescope, with many modes of operation. With some new optics it has good images over the wide field of 2° (four times the apparent diameter of the Moon), and this has been exploited in the same way as with the AAO Schmidt telescope by using an optical-fibre system which feeds light from four hundred different objects into spectrographs mounted on the observatory floor. Instead of measuring only one spectrum at a time, this enables a large survey of distant high-redshift galaxies to be undertaken in a reasonably short time. The equatorial mount of the AAT has an advantage for this multiple-spectrum observation: the image of the sky, with all four hundred optical fibres, moves with the telescope without rotating. Nevertheless, the engineering difficulties in using the equatorial mount for a next generation of even larger telescopes were formidable, and in all new designs it was abandoned in favour of the altazimuth mount.

3

NEW WAYS TO BUILD BIG TELESCOPES

The big Californian telescopes and their immediate successors opened up a new window on the Universe. Our own galaxy, the Milky Way, was seen as a spiral nebula, with stars following a sequence from a collapsing gas cloud through evolutionary tracks to eventual death by fading, collapse, or explosion. It was now evident that similar processes were at work in billions of distant extragalactic nebulae; furthermore, these galaxies could be observed to such large distances that the evolution of the whole Universe could become a subject for meaningful research. Cosmology, like the Universe itself, was a science that was rapidly expanding. Another new generation of even bigger telescopes was needed, and fortunately there were some new ideas on how they could be built.

The design objectives of the new generation were first to collect enough light from distant faint objects to determine their structure and composition from their spectra, and second, to repeat this for a sufficiently large number of distant galaxies to reveal the structure and evolution of the Universe. The largest possible telescopes would be needed, with the capability of simultaneously observing over a large field of view. The volume of data collected by a single telescope threatened to be overwhelming. Large-scale computation and data storage would be essential, but fortunately the digital age was developing at just the right time. The first new development was in the engineering field of telescope mounts.

The Altazimuth Mount and Computer Control

It is hard to image life without digital computers, but before the Anglo-Australian Telescope came into operation in 1974 the only telescopes moving under computer control were some radio telescopes with altazimuth mounts. Following an object as it moves across the sky needs movement in both elevation and azimuth, and for the altazimuth mount there has to be a continuous computation of position to convert a celestial position to the required telescope position. The 250-foot Mk I radio telescope at Jodrell Bank originally, in 1957, used an analogue computer for this coordinate transformation, and the 210-foot Parkes radio telescope was originally linked optically to a tiny but precise polar-mounted mechanical model. The Mk II telescope at Jodrell Bank was the first telescope of any kind to be controlled by a digital computer, in 1964. Applying this idea to an optical telescope may have upset some conventional astronomers and telescope designers, but it was an essential step in the development of the next generation of large optical telescopes.

The possibilities of an altazimuth mount for a large optical telescope were explored in the Russian BTA-6, the 6-metre telescope built in 1975 at the Special Astrophysical Observatory in the Caucasus. The principle was clearly demonstrated, and the altazimuth mount was adopted by all large telescopes from then on. The BTA-6 telescope itself was less successful, and served to provide lessons for future designers. The site of the telescope unfortunately proved to be poor, both in weather conditions and atmospheric turbulence; but the main problem was in the 6-metre mirror itself. The original mirror had surface cracks and other defects, and in 1978 it had to be replaced. The difficulties of making such a large mirror, from the casting of many tonnes of ceramic material to figuring and polishing the surface, then transporting it to a distant mountain site, and the engineering of mounting such a heavy mirror stably within the telescope, were obviously becoming almost impossible to overcome. Furthermore, these large and thick mirrors are sensitive to temperature variations, even for low-expansion materials, and some distortion is inevitable.

Thin Mirrors

The experiences of the USSR 6-metre telescope demonstrated that the possibilities for even larger telescopes depended on a new approach to making large mirrors. Scaling up the already massive mirrors of the 5- and 6-metre telescopes faced huge problems of casting, cooling, figuring, and mounting, with continued problems of temperature control in operation. Thinner mirrors might be more practical, but the problem was that thin mirrors bend under gravity, and the conventional mirror-support systems could not control the flexing which seemed inevitable when the telescope moved to different elevations. A new support system was needed that would actively compensate for changing gravitational forces. The development of such a system became part of an investigation of new technology by the European Southern Observatory (ESO).

ESO is a consortium of European observatories combining their resources in observational astronomy, set up in 1962 with the primary objective of building large telescopes on a first-class site in the southern hemisphere. In 1979 ESO started the design of a 4-metre class telescope as a test-bed for new ideas, which would lead to a design for a Very Large Telescope (VLT) following soon after, and eventually to an Extremely Large Telescope (ELT). The test-bed telescope, which became an important instrument in its own right, is the 3.8-metre New Technology Telescope (NTT), completed in 1989. As well as using an altazimuth mount, the NTT has a thin mirror only 24 cm thick. This meniscus of glass-ceramic weighs only a quarter as much as the mirrors of several other contemporary 4-metre class telescopes. Such a thin mirror tends to distort unacceptably under gravity as the telescope moves in elevation, but an active system of supports continually adjusts for the changing forces and preserves the profile. This was the first example of 'active optics', which is now adopted in all large telescope designs.

Fig. 17. The New Technology Telescope (NTT) at La Silla, Chile. The wide opening of the sliding roof avoids 'bad seeing' due to thermal fluctuations in the telescope enclosure.

The NTT, which is on La Silla—an excellent site at an elevation of 2,375 metres in Chile—produces stellar images often less than 1 arcsecond in diameter. Among the new ideas tested at the NTT was an innovative design for the enclosure, replacing the conventional dome (Figure 17). The small focal ratio, and the Gregorian secondary, allowed a compact telescope to be fitted into a comparatively small hexagonal cylinder with a sliding roof. This was very economical in comparison with the traditional spherical dome, and it also allowed an extensive system of ventilation through louvres in the sides. Air circulation is very important to avoid thermal disturbances in the path of the light, especially close to the mirror. Astronomers refer to the atmospheric effects causing fuzzy star images as 'bad seeing'. Previous dome designs had led to bad dome seeing, which can be avoided only by strict temperature control.

An outstanding example of the imaging capabilities of the NTT was an investigation of the centre of the Milky Way galaxy. This was done in infrared light, which is less absorbed by interstellar clouds than is visible light (see Chapter 4). It led to the discovery of a group of stars in orbit round the black hole at the Galactic centre, though a full resolution of these stars and their orbits eventually required a herculean effort using the next generation of telescopes. Most of this group of stars are contained within an area only 1 arcsecond across.

The VLT, which followed the NTT, used the same ideas of altazimuth mounting and a thin mirror primary, but pushed the mirror diameter up to 8.2 metres—almost the largest ever to be cast all in one piece. (The largest, 8.4 metres in diameter, have been cast at the Steward Observatory Mirror Laboratory of the University of Arizona.) The VLT comprises no fewer than four such telescopes, located close together on a single site (Plate 3) so as to allow joint operation as an interferometer (see Chapter 9). The first was completed in 1998, and all four were in operation by 2000. The site, Cerro Paranal, is again in Chile, at an elevation of 2,635 metres. A similar telescope, Subaru (the Japanese name of the well-known Pleiades star cluster), also 8.2 metres in diameter, was built by the National Astronomical Observatory of Japan, and was completed on Mauna Kea, Hawaii, in 1999. These telescopes clearly represented the culmination of a line of development of large mirrors. Thinning to 20 cm or less, and using active optics to compensate for the force of gravity, had allowed the diameter to reach this dramatic size, but handling and transporting any larger single piece of glass was evidently becoming impossible. The solution was the segmented mirror, using easily handled sectors only 1 metre or so across, individually figured and closely fitted together with very precise controls to make a single reflecting surface.

A further problem with these monster telescopes is to build domes large enough to contain them and protect from the wind and weather. The overall length of the telescope determines the dome size, so that while designs using larger and larger-diameter mirrors were proposed, the focal length and with it the curvature of the mirrors remained the same. It is also helpful to use a Gregorian design, in which a secondary mirror reflects the light back through a central hole in the main mirror, or, as in Nasmyth's design, out to one side through a hollow bearing. The use of a secondary mirror actually offers an optical design advantage, since the optical designer may now choose a shape for both the primary and the secondary to produce optimum overall performance. A combination of two hyperboloids rather than paraboloids, the Ritchey–Chretien or RC design, which produces better image quality over a wide field of view, was adopted for the VLT.

Segmented Mirrors

The segmented-mirror concept was proposed in 1977 by Jerry Nelson, then at the Lawrence Berkeley National Laboratory,[1] during the early design phase of the Keck telescope, which was to be built by a consortium of Californian institutes. The initial design concept for this telescope was to push the size of a single mirror to 10 metres, but in 1979 Nelson persuaded the design team to investigate techniques for mounting thirty-six thin hexagonal mirrors, each 1.8 metres across and only 7.5 cm thick, to form a single perfect hyperbolic reflector surface. The difficulty was to figure individual elements to be part of an overall hyperboloid; grinding and polishing a symmetrical spherical surface was routine, but the required surface of a segment of a hyperboloid is asymmetrical. The solution was to stress each thin mirror in a special frame, distorting it by a calculated amount, and polish the warped mirror to a spherical shape. It should then relax into the required shape. A problem arose in cutting the edges to fit precisely together; cutting relieved small remaining stresses which distorted the edges, and this had to be compensated by applying pressure at the edges of each mirror when it was mounted in its cell. The separate segments must be aligned to make a continuous surface. This alignment is achieved by actuators acting on the supports of each segment, which follow differences in alignment measured by sensors at the mirror edges. The alignment is checked and corrected twice each second.

The final product was a great success; the whole surface of the mirror is accurate to a small fraction of the wavelength of visible light. The Keck telescope, built on the mountaintop of Mauna Kea (Plate 4) and completed in 1993, was followed by an identical telescope, Keck II, which was built on an adjacent site and completed in 1996. The two can be connected to act as an interferometer (see Chapter 9).

Segmented mirrors, with hexagonal elements around 1 metre across, have been adopted in all new large telescopes, with one exception. Roger Angel, at the Steward Observatory Mirror Laboratory, developed an efficient process for casting 8.4-metre thin mirrors (Figure 18). The Laboratory is now engaged in the production of seven of these which will be mounted together to form a single telescope 25 metres across. (Allowing for gaps between the circular elements, the collecting area will be equivalent to a 22-metre aperture.) This will be the Giant Magellan Telescope, a joint US enterprise led by the Carnegie Institute, with partners in Australia and

Fig. 18. An 8.4-metre mirror ready for polishing at the Steward Observatory Optical Laboratory.

Brazil. The individual mirrors are cast in a slowly revolving mould, in which centrifugal force combines with gravity to produce a parabolic surface profile. All seven mirrors must then be ground into a single curved shape, which for this telescope will be part of an ellipsoid. The outer six off-axis segments require grinding and polishing to an asymmetric profile.

Mounting these huge telescopes so as to track smoothly across the whole sky is a major engineering problem. Some relief is offered by using an altazimuth mount with only limited sky coverage; an example is the 9.2-metre Hobby–Eberly telescope built by the University of Texas at their McDonald Observatory in 1997, followed by an 11-metre version at the South African Astronomical Observatory. These telescopes, surprisingly, operate only at a fixed angle above the horizon; the nearly complete sky coverage is achieved by movement in azimuth only, to a direction which will be crossed by the target point in the sky. The telescope then waits for the target to achieve the fixed altitude in the sky. During an observation the telescope does not move, but the image of the target moves across the focal plane. If the telescope were to use a photographic plate, that would have to be moved along with the image; but instead, a stationary large CCD array is used, and the image is tracked electronically across the individual detectors of the array, the outputs of which are added in a precomputed programme. The main mirror is segmented, comprising ninety-one identical hexagonal mirrors which together form a surface with spherical rather than a parabolic symmetry. Correct focusing requires a specially figured secondary mirror. This is closely analogous to the design of the Arecibo radio telescope, which also has a spherical reflector surface (see Chapter 8).

Incidentally, the naming of many of these telescopes after major donors who financed them follows the pattern of the sequence of great Californian telescopes: the Hale 60-inch and the Hooker 100-inch on Mount Wilson, and the 48-inch Oschin Schmidt on Mount Palomar. W. M. Keck is unique in having two 8-metre telescopes named after him. Funding the next generation, which may cost in the order of a billion dollars (or Euros), is less likely to follow this pattern, and will have to be an international responsibility.

Sharpening the Image

Every physics student knows that the sharpest image available from a telescope is determined by the ratio between the wavelength of light and the diameter of the telescope aperture. An aperture of only 20 cm should ideally produce images as sharp as 1 arcsecond, which would show detail on the Moon as small as 2 km across (the diameter of the Moon is 3,470 km). The current generation of large telescopes should, according to this simple rule, easily achieve a resolution of better than 0.1 arcsecond. Unfortunately, the terrestrial atmosphere upsets this simple calculation. Light may travel for thousands or millions of light-years across the space between the stars without appreciable distortion, but when it reaches the turbulent air in our atmosphere it is scattered and blurred so that no star appears to be less than about 1 arcsecond across. The effect is variable; at the best mountaintop observatories the seeing may often be much better, but never as good as the large telescope diameters should provide.

Star images affected by this atmospheric scintillation often appear as smaller images, nearly as small as the ideal but moving rapidly in a random motion within a larger disc. Short photographic exposures may produce useful images with good resolution, but the longer exposures needed for faint objects are blurred to the full size of this scintillation disc. Following the image as it dances around requires the telescope to chase after the image with a response time of less than a tenth of a second. Moving the whole telescope at that rate is impossible, but a small lightweight mirror near the focus can be moved sufficiently rapidly to keep up with the image. This active mirror must be responding to an even faster measurement of the image movement, which is most easily achieved by projecting a bright star image within the field of view onto the centre of a small square array of four detectors. A central star image produces equal signal in all four, so the ratio of their outputs is used to control the angle of a small mirror on two axes, usually known as tip and tilt. There must be enough light at the detector for this to be achieved at least ten times per second, so it is only possible for a star which is bright enough to produce a large signal in such a short time. Fortunately, the image movement is the same over an appreciable field of view, so the trick is to track the movement of a bright star near the faint target and use that measurement to control the 'tip and tilt' active mirror. This system having been shown to work, the way was open for a more comprehensive active optics system, known as adaptive optics, which has transformed the performance of large telescopes.

Scintillation is more complicated than a simple movement of a star image; the image is blurred in a randomly changing pattern. Ripples of light on the bottom of a swimming pool, generated by a random pattern of waves on the water surface, have a similar effect. In the telescope we may think of an ideal lightwave with a plane wavefront reaching the atmosphere and emerging corrugated in a pattern of waves around 10 cm across, which changes randomly on a time-scale of less than a tenth of a second. Straightening out this pattern requires more than a tip–tilt mirror; it requires a mirror that can be distorted with a pattern which is the opposite of the corrugations in the wavefront. The distortions are small—generally no more than a wavelength of light—and can be achieved by stressing a thin mirror with some tens of rapidly acting pressure pads in its support system.

The corrugations in the wavefront are measured when it is converging towards the telescope focus, and are only a few centimetres across. A tightly packed array of tiny telescopes makes an array of images of the bright star, each using only a small area of the wavefront. The slope of the wavefront in each area shifts the image, and this shift is measured for each in a small detector array. The corrugations can then be computed and the corrections transmitted to the active mirror. As for the simple tip–tilt system, a bright star is needed as a reference, which must be fairly close to the target area; an obvious application is to search for a faint planet close to a bright star. This is a severe restriction, because the light from the bright star has to be divided between the many tiny telescopes, each with its own movement sensor. Fortunately, the need for a bright star near the target area is no longer a restriction, since it is standard practice to produce an artificial bright star by shining a laser beam which creates a bright spot in the upper atmosphere. This system of adaptive optics should be distinguished from the simpler 'active optics', which is preprogrammed to preserve the shape of mirrors which would otherwise distort due to gravitational forces. Adaptive optics now enables all large telescopes to produce images with ideal resolution anywhere over the whole sky.

Adaptive optics has made possible some amazing new observations. Plate 5 is an image of Uranus obtained with the Keck Telescope on Mauna Kea, using adaptive optics. The detail of the ring and the surface features could not be seen in ordinary ground-based observations.

One of the four telescopes of the VLT, using the artificial laser star as reference, achieves a resolution of 50 milliarcseconds. The concentration of light in such a tiny image enables the detection of very faint objects (to

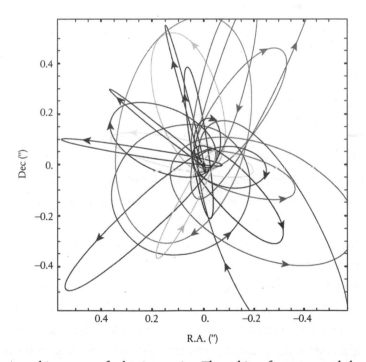

Fig. 19. An achievement of adaptive optics. The orbits of stars around the central black hole of the galaxy have been measured using adaptive optics on the VLT and Subaru 8-metre telescopes. The whole of this map fits inside a square only 1 arcsecond across.

magnitude 30 in astronomers' terminology, or four billion times fainter than can be seen by the human eye). An example at the VLT was the first actual image of an extrasolar planet, orbiting round the star β Pictoris.[2]

Adaptive optics installed on the VLT and Subaru has achieved the astonishing feat of resolving the group of stars surrounding the central black hole of the Milky Way galaxy. Combining observations at these two telescopes, astronomers have produced a detailed map[3] of an area only 1 arcsecond across (Figure 19). More than twenty individual stars have been individually resolved within this tiny area, and their orbits tracked in successive observations; they are in orbits with periods of only a few years. The dynamics of these stars gives a measurement of the mass of the black hole as four million times that of the Sun.

The Big Surveys

If the whole sky were to be mapped with a resolution of 1 arcsecond, there would be 50,000 megapixels (5×10^{11} pixels) in the map. Before the advent of digital computers, and *a fortiori* massive digital storage, the only way of even approaching such a massive task was via photography: the Carte du Ciel and the Schmidt telescope photographic surveys are the outstanding examples, whose collections of photographic plates are still used as an archive for comparison with more recent observations of variable stars. Even the data on these plates are now stored digitally, after painstaking scanning with specially built machines. The situation was transformed by the use of CCDs, arrays of solid-state electronic detectors, matched to specially designed telescopes. The outstanding example is the Sloan Digital Sky Survey (SDSS, Plate 6), which started observing in 1998, with full surveys beginning in 2000.

The SDSS project, initiated by Princeton University under the leadership of James Gunn, has graduated into an international project largely because of the enormous amount of data it produces. The 2.8-metre telescope is a straightforward two-mirror design, with the camera at its Cassegrain focus (behind the main mirror) and two thin correcting lenses near the focal plane. The field of view is 1.5°, which is covered in the main survey by five CCD arrays, each covering a colour band of the spectrum. The whole array, comprising 120 million individual detectors, is cooled by liquid nitrogen at −80° C. Unusually, but in a similar mode to the Hobby–Eberly telescope, the telescope is kept stationary during an observation, so that the image of the sky drifts across the five detector arrays and is reconstructed in the computer as a stationary image in each of the five colours. The initial survey (2000–05) with the SDSS covered one fifth of the sky, producing five-colour photometry on 500 million objects. Many of these are galaxies, the redshifts of which can be estimated from the five-colour spectra.

The SDSS also operates in a fully spectroscopic mode, using fibre optics to connect up to 1,000 objects to stationary spectrographs. In contrast to the system used in the AAT and Schmidt telescopes at the AAO, the objects are selected by drilling holes in an aluminium plate at the focal plane, locating the optical fibres for each target galaxy. In this mode of operation the telescope has to track the sky, while the spectrograph outputs are detected in CCD arrays. More than 1 million galaxy redshifts have been measured—some as large as $z \approx 0.7$ (redshift is the proportional increase of wavelength which gives a measure of distance). Quasars—the exceptionally bright cores of some galaxies—have been found with redshifts up to $z \approx 5$.

The possibilities of very-large-scale surveys opened up by the combination of multimegapixel detectors and the computer power to handle the huge data-flow that they can produce has inspired the equipment of several telescopes for wide-angle observations. The AAT adaptation for a wide field has been followed by a similar adaptation in another almost contemporary 4-metre telescope: the Victor Blanco telescope at Cerro Tololo, Chile. Here, as in the AAT, corrector lenses have been added to expand the field to a diameter of 2.2°, with a 570-megapixel array of detectors. The main objective is to explore the dynamics of the Universe on a very large scale. This is believed to be dominated by 'Dark Energy'—a force tending towards expansion and opposing the gravitational tendency towards collapse—of the Universe. The CCD camera is accordingly named the

Dark Energy Camera (DECam). Other telescopes for similar purposes have been designed primarily for operation at infrared wavelengths. These are described in Chapter 4.

The Next Generation

There is already under construction a new survey telescope which combines a wide field of view with excellent image quality and great sensitivity, providing an unmatched capability for rapid and repeated surveys of the majority of the sky. This is the 8.4-metre diameter Large Synoptic Survey Telescope (LSST), expected to be in operation in Chile in 2020 or 2021.

The demanding specification of the LSST is met by a three-mirror configuration,[4] as shown in Figure 20. This configuration results in a telescope which is more compact than the Schmidt, its predecessor in large sky surveys. The startling feature is the huge third mirror, located in the same plane as the primary and occupying no less than 5 metres of the total 8.4-metre aperture. These two mirrors are combined in the same piece of glass. Three further optical elements are mounted in front of the camera; these are

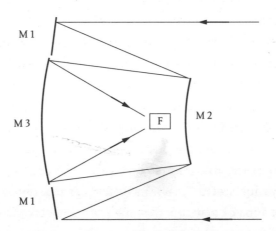

Fig. 20. The optical design of the 8.4-metre Large Synoptic Survey Telescope. The third mirror, 5 metres in diameter, occupies the centre of the primary 8.4-metre mirror, and is part of the same piece of glass.

low power lenses, giving a corrected field of view 3.5°. The image quality over this very large field is so good that it requires a camera with 3.2 billion pixels, the largest in any telescope.

The LSST will be able to repeat surveys of the whole sky at intervals of only three days, picking up transients such as the very distant supernovae, which are important indicators of cosmological geometry. The data rate expected from the LSST is phenomenal. The camera will be read out every 20 seconds, requiring a huge computer power (100 teraflops) and a huge storage capacity (15 petabytes). Handling such a flow of data, and mining it for useful information, is perhaps the most challenging task for the designers of this observatory.

Two new very large telescopes, taking the story of increasing size to another order of magnitude, are under construction, one in each hemisphere. The motivation is to build the largest possible collecting area, giving the greatest possible sensitivity for observing faint objects, such as exoplanets and the most distant galaxies in the early Universe. The Thirty Metre Telescope (TMT), which will be on Mauna Kea, Hawaii, is a joint effort between the USA and Canada, with contributions from China, Japan, and India. The main mirror will have 492 hexagonal segments. The European Southern Observatory is constructing the E-ELT, an Extremely Large Telescope 39 metres in diameter (Figure 21). This will be situated on another mountain peak in Chile: Cerro Amazones, in the south of the Atacama Desert. The main mirror of the E-ELT will comprise no fewer than 798 individual hexagonal segments, each of which is 1.45 metres across and might itself be regarded as a fairly large mirror. The overall ratio of thickness-to-diameter in this huge primary mirror is only 1 percent of that in the 200-inch (5-metre) Hale telescope, as seen in Figure 22.

Producing the component mirrors of these telescopes, at the rate of one every few days for several years, including casting, figuring and optical testing, will be a major new task for the optical industry. The scale of these instruments is hard to comprehend; even the secondary mirror of the E-ELT is 4 metres in diameter—the size of the largest telescope mirrors being built only half a century earlier. The adaptive optics systems will be on a

Fig. 21. The ESO Extremely Large Telescope, under construction at Cerro Armazones in Chile. For scale, note the two people on the Nasmyth platform.

huge scale; the corrector plate of the E-ELT will be a thin mirror 2.4 metres in diameter. It will require 8,000 actuators, all operating with a response time of 1 millisecond. Four lasers will operate simultaneously to provide artificial stars over an extended field of view.

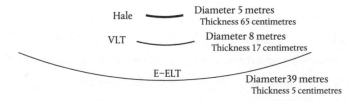

Hale — Diameter 5 metres
Thickness 65 centimetres

VLT — Diameter 8 metres
Thickness 17 centimetres

E-ELT — Diameter 39 metres
Thickness 5 centimetres

Fig. 22. Mirror thickness. This scaled diagram shows how the ratio of thickness-to-diameter has changed, allowing the construction of the new extremely large telescopes.

Designing the E-ELT, which will cost over 1 billion euros and take ten years to build, requires an interesting balance between innovation and caution. Adaptive optics will be an integral part of the design, but the huge scale is without precedent; the possibility of incorporating tip–tilt and the more rapid scintillation compensation, along with the active optics which compensates for gravitational distortion, all in a single mirror, is regarded as complex, demanding, and even somewhat risky. There will be five mirrors, allowing the different functions to be separated and also allowing a design for a wide field without transmission through lenses. 'Wide field' for such huge telescopes is measured in arcminutes rather than degrees; there is no prospect of scanning the whole sky, which is the domain of smaller survey telescopes.

The E-ELT will be protected from the weather, and from the heat of the Sun, by a fairly conventional dome. The dimensions of domes for the 4-metre class telescopes were determined mainly by the length of the telescope itself, while the new designs have moved towards increased diameter without a proportionate increase in focal length—that is, towards smaller f-numbers. With a focal ratio of 0.93, the E-ELT will be wider than it is long; the Nasmyth platforms, where the cameras and spectrographs will be mounted, extend the width even further, and the overall structure will be 86 metres wide. The control of temperature in several of the domes for the 4-metre class telescopes has proved to be inadequate, leading to random refraction in the air inside the dome. The solution for the E-ELT is to allow free circulation of outside air through the whole dome.

What will the new generation of huge optical telescopes achieve? An increase in sensitivity by a factor of ten or more over existing telescopes will open new fields which we cannot imagine, but there are some pressing issues which may at least be addressed, if not resolved, by the E-ELT and its like. The increase in sensitivity will take observations of galaxies closer to the early cosmological era when the complex structures of clusters and individual galaxies were developing. For many astronomers, and the wider public, the most intriguing subject is the question of planets around stars other than our Sun, and the physical conditions which might support life on them. The existence of these extrasolar planets is already well established; in fact, some hundreds have been detected, either by measuring the tiny diminution of light from a star as a planet crosses its disc, or by observing the effect of the gravitational pull of a planet as it orbits round its star. What is so difficult is to observe the planet itself, since light from the star is overwhelmingly bright. A very high angular resolution is needed to separate the images of the star and the planet, but in addition it is essential to prevent the starlight from scattering over the image plane. Precision optics, a minimum of structure in the optical paths, and a high dynamic range in the detector system will be needed. Compensating for atmospheric disturbances will be the key to success, though it remains to be seen whether this can be achieved sufficiently well to compete with telescopes operating in the clear environment of space.

4

STRETCHING THE SPECTRUM: INFRARED AND ULTRAVIOLET TELESCOPES

In 1666 Isaac Newton split a beam of white light from the Sun into a spectrum of colours. He used a prism, and later used a second prism in reverse to reconstruct white light from the spectrum. The nature of light, as a wave, was not established until more than a century later, by Thomas Young (1773–1829) and Augustin-Jean Fresnel (1788–1827). The wavelength was found to be a fraction of a millionth of a metre, and it then became clear that the separate components of white light were defined by their wavelengths, which ranged over a factor of nearly two across the visible spectrum. Modern terminology uses as a unit the nanometre (10^{-9} metre), abbreviated to nm; the wavelength range of visible light is now reckoned to be from 380 to 750 nm, covering the violet end to the red end of the spectrum. No-one seems to have speculated whether there was any form of radiation outside the visible range, until in 1800 William Herschel explored the effect of light of different colours on a thermometer placed in a spectrum of sunlight. To his surprise, the largest heating effect was beyond the extreme red end, outside the visible spectrum. He had discovered infrared radiation. The extension to the short wavelength violet end was discovered a year later by Johann Wilhelm Ritter (1776–1801), a chemist who noticed the effect of ultraviolet light on silver chloride.

The rods and cones in our eyes have evolved to cover most of the range of wavelengths in light from the Sun. This visible spectrum lies within the range of wavelengths, or colours, for which the terrestrial atmosphere is transparent (when it is not cloudy). As Herschel discovered, there is also infrared radiation reaching us from the Sun, which behaves in many ways like visible light. Cosmic infrared radiation, at the long-wavelength end of the radiation that reaches the surface of the Earth, carries unique information about stars and about molecules in the interstellar medium. The available wavelength range of the infrared is large, reaching more than ten times the red end of the visible spectrum. The ultraviolet light which reaches the Earth's surface has a smaller extension from the visible, by just over a factor of two in wavelength; it has its own domain in astrophysics, related more to atoms than to molecules. Conventional telescope techniques can be extended in both directions from the visible spectrum, but it is the infrared that offers the most reward and presents the major challenge. The key to successful infrared telescopes lies in the detector systems.

Photons and Waves

The difficulty of detecting infrared radiation lies in a fundamental property of all radiation. All electromagnetic radiation, including light, has a dual nature: wave and photon. The waves are oscillating electric and magnetic fields, but the energy in a wave exists in discrete amounts or quanta, the photons. The relation between the wave and the photon is an enduring mystery in physics. From the astronomer's point of view we can say that a wave arrives at the telescope aperture and is focused via mirrors and lenses onto a detector. At this stage we change our viewpoint and our terminology, and regard the incoming radiation as a shower of photons which arrive on the detector and transfer their energy as individuals. The photon energy depends on the wavelength. For a photon to be detected by our eyes or by a photographic plate, or by the electronic detectors in a modern camera, it has to have sufficient energy. If the photon has sufficient energy to overcome a barrier in a detector, it can be recorded as an electrical impulse, or as a chemical change in a photographic plate. Below this critical level, a photon will go unrecorded.

The energy in the photons of long-wavelength infrared radiation is not enough for them to be detected in the usual detector systems used in the visible spectrum, including the photochemical receptors in our eyes. Infrared astronomy has become a major growth area only because of the successful development of new ideas in sensor technology, which allows low-energy photons to be detected with high efficiency.

The term 'infrared' applies to a wide range of wavelengths and photon energies (Figure 23). The wavelengths used in terrestrial and space telescopes cover a range of more than a hundred, starting at the red end of the visible spectrum. These long wavelengths are usually specified in units which appropriately are stepped up by a factor of 1,000 from the optical units: instead of nanometres (10^{-9} metre) we use micrometres (10^{-6} metre), abbreviated to microns, and the infrared range is reckoned to spread from 0.75 to 300 microns. Using the new detector technologies, the shorter infrared wavelengths, up to about 5 microns, can now be treated much like visible light. This range is called the near-infrared. With some effort, detector systems can be made to work up to 10 microns, which is often quoted as the limit of the near-infrared.

For the longest infrared the photon energies are so small that they cannot be detected as individuals by any detector systems. Instead we must follow Herschel in using some kind of average energy detector, like his thermometer but very much more sensitive. The first attempts used a thermocouple, or a thermopile (a stack of thermocouples), in which the voltage at a junction between two metals depends on temperature. The faint heating effect of infrared radiation from the Moon and some stars was

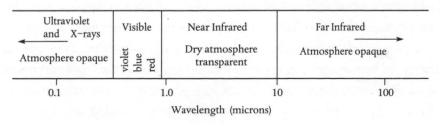

Fig. 23. Observations from the ground are possible within the near-infrared, which covers a decade of the electromagnetic spectrum.

detected in this way, but sufficient sensitivity seemed to be out of reach, and it was not until the mid-twentieth century that serious infrared astronomy could start. As in the visible, the key to success in the near-infrared is in assembling a large array of individual sensors to act like the CCD, which is at the heart of all modern cameras.

We now need a simple explanation of the way in which solid-state physics has allowed these remarkable developments. The world of quantum physics is not familiar in everyday life, and some new concepts and terminology must be introduced.

Photons and Electrons

The simplest concept in a photon detector is that of a photoconductive diode. Silicon is a semiconductor; that is, it is normally an insulator but can conduct when suitably stimulated. If a cube of pure crystalline silicon has an electric voltage applied across it through electrodes on two opposite faces, no current flows. The electrons in the crystal are locked and cannot move unless they are given enough energy to overcome an energy barrier known as the band gap. They can then move, and a current flows which can be amplified and detected. The free electron may alternatively be stored as a charge within the detector diode, and the charge accumulated from several photon events can be read out after some time. Even a single electron can be detected in the tiny diode elements of modern detectors.

The size of the band gap is specific to the material; for silicon it is 1.1 eV. For a photon to be detected it must produce an electron of at least this amount of energy. This determines the wavelength limit for photon detection, which means that a pure silicon photoconductor diode cannot detect the low-energy photons of infrared light. Materials with a smaller band gap which are currently in use include exotic compounds such as indium antinomide (InSb) and mercury cadmium telluride (HgCdTe), which work over most of the near-infrared. Materials with smaller band gaps, which can extend the wavelength range, can be made by introducing impurities into the pure crystalline materials. Unfortunately, a smaller gap also means that electrons may have enough energy to surmount the band gap without any outside stimulation; this is an intrinsic effect which cannot be avoided and which depends on temperature. Even for semiconductors operating in visible light and at normal room temperature, there is an unavoidable so-called 'dark current' originating in thermal excitation of electrons. Many amateur astronomers cool their electronic detectors to reduce this dark current so as to obtain the best performance for faint objects. All infrared telescopes now use detectors cooled to liquid nitrogen temperatures (−196° C, or 77K, degrees absolute) for the near-infrared, while for longer wavelengths detectors may require extremely low temperatures, using liquid helium (at 4K) as a coolant. Detectors are routinely cooled by mechanical refrigerators. The days are long past when each night, at the start of observing, astronomers at some telescopes had to pour liquid nitrogen into their CCD cameras to cool them. Many modern systems have refrigerators which routinely reach down to 1 degree above absolute zero (1 K).

Arrays and CCDs

A single infrared sensor, however sensitive, is not very useful in astronomy. The electronic digital camera has only taken over from photographic film because it uses arrays of millions of sensors, each detecting one element or pixel of the picture, and the quality of an image is often determined by the number of megapixels in the camera. Each sensor element may be only a few microns across, so that an array of a million sensors may be packed into 1 cm square. Making an electric connection to all these individually, allowing them all to be read simultaneously, is impossible. Instead they are read serially, using another vital element of digital cameras: the CCD.

A CCD reads in series the signals from an array of individual detectors, each contributing a single pixel in a photograph. A photon detected in a single detector creates a small electric charge, and the CCD reads these one

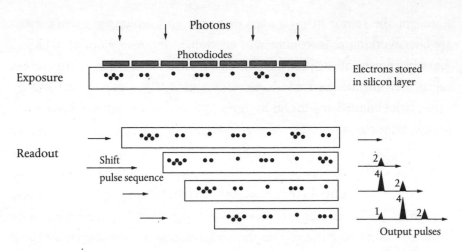

Fig. 24. The charge-coupled device (CCD). Electrons from an array of photodiodes accumulate in a silicon layer behind the diodes. They then move in steps along the array, in response to a series of shift pulses, prducing an output sequence in a single electronic circuit.

by one, so that the image becomes a long digital signal which can be used to construct the whole picture (Figure 24). The array is a mosaic with thousands or millions of units, and even the cameras in mobile phones have many millions (many megapixels). The whole array is etched, deposited, or printed onto a single thin slab of crystal, usually of silicon. The pixel elements are arranged in rows. Reading a single row has been described as like a bucket brigade passing buckets of water down a line; the electric charge on each diode is passed down the row and emerges in a single amplifier at the end of the row. The rows are then read out in series, to be stored digitally or to appear on a display screen.

For visible light the unit detectors are semiconductor diodes that are part of the CCD itself. This works well in ordinary digital cameras, and also works so efficiently in telescopes that 90 percent or more of photons in the faintest of light signals can be detected. The same applies in the near-infrared, but the mid-infrared and far-infrared require different detectors which operate only at very low temperatures. In the extreme example of the Transition Edge Sensor (TES), the sensor must work at only 1 degree above absolute zero. The detector array then presents a formidable challenge in design; the sensor in each pixel element must incorporate a separate amplifier working at low temperature, coupled to an element of a CCD. The achievement of such an array has opened up the far-infrared to wavelengths beyond 100 microns (0.1 mm). There is now no inaccessible gap between the infrared and the shortest part of the radio spectrum, known as the submillimetre.

The Troublesome Atmosphere

There are only two parts of the electromagnetic spectrum for which the Earth's atmosphere is completely transparent: the visible and radio regions. The visible region extends into the near-infrared, but even in that region there is absorption in the molecular constituents of the atmosphere, and especially water vapour. Beyond a wavelength of about 30 microns there is little hope of doing useful astronomy from the ground, but in the near-infrared the absorption by water vapour is concentrated in a number of characteristic spectral lines. Useful observations can be made only by using filters that, so to speak, allow us to look through windows between the absorption bands. Since the main cause of the absorption is water vapour, the best observatory sites are at high altitude, above the clouds. The best conditions that can be hoped for in this part of the spectrum are outlined in Figure 25, which shows the transparency of the atmosphere varying between 0 and 100 percent across part of the infrared. Water vapour is the main contributor to the forest of absorption lines. This graph applies to the minimum amount of water vapour encountered at the mountaintop of Mauna Kea in Hawaii—one of the best sites in the world.

Absorption of infrared is not the only problem. The atmosphere itself radiates in the infrared—most strongly at wavelengths where it absorbs most strongly. This thermal radiation is an inevitable background to observations, even in the clearest bands seen in Figure 25. It is a very serious barrier to observations in the broad band from 8 to 14 microns, but within the bands between 1.1 and 4.0 microns there are opportunities which have

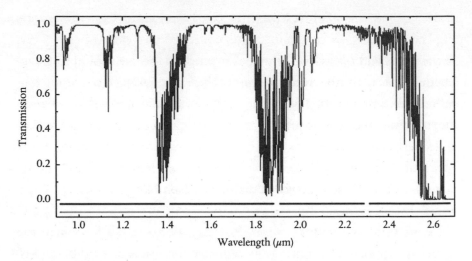

Fig. 25. Atmospheric absorption in the near-infrared for a mountaintop observatory with minimum water vapour. Observations are best made in restricted bands, such as 2.1–2.3 microns, avoiding the multiple atmospheric absorption lines.

been, and continue to be, exploited in the design of large and powerful telescopes.

Thermal effects are generally a serious problem for infrared telescopes; not only does any warm detector itself produce unwanted signals, but also the mirror and the structure of the telescope radiate some stray thermal radiation. The combination of random signals from the detector, the telescope structure, and the atmosphere is an unwanted signal known to astronomers as thermal background noise. Thermal radiation from the structure of a telescope is a serious problem. At ordinary room temperatures the peak of thermal radiation is in the infrared; an infrared photograph of the outside of a house shows it glowing with a brightness that reveals the temperature of the walls and roof. Cooling the whole structure of a telescope is, however, impractical. The design aim must be to arrange that as little as possible of the thermal radiation from the structure can enter the detector system. Any structure in front of the detector must be as sparse as possible, making it a lightweight web rather than a solid tube. Most of the telescope structure has to be 'out of sight' of the detectors.

Telescopes for the Infrared

In the 1970s and 1980s, two telescopes were built on the mountaintop of Mauna Kea with the intention of exploring the infrared and sub-millimetre wavebands. There is a large gap between these wavebands, due to atmospheric absorption; no ground-based observations are possible between wavelengths of 30 and 450 microns. The James Clerk Maxwell Telescope (JCMT), for the longer wavelengths of millimetre-wave astronomy, is 15 metres in diameter, resembling a radio telescope, while the UK Infrared Telescope (UKIRT) follows optical designs, with a 3.8-metre glass-ceramic mirror. The link between the two is in the detector systems: both must explore new types of detector working at very low temperatures, and both must take account of thermal radiation.

UKIRT was constructed throughout 1975–1978, and started observations in 1979. It was designed to observe the near-infrared, intending to reach as far as 30 microns. Its construction largely followed principles established for conventional optical telescopes, using an equatorial mount, but with the innovation of a thinner, lightweight primary mirror. The development of detector arrays came after the telescope was built, and it was not until 2004 that UKIRT started on its most productive era, when it was equipped with a massive camera using four CCD arrays, each of 2,048 x 2,048 pixels. The objective was to survey a large proportion of the sky in the near-infrared; the example for this had been set by 2MASS.[1]—a survey of the whole sky during 1997–2001, using two identical dedicated 1.3-metre telescopes in the northern and southern hemispheres. The 2M in the acronym denotes 'two micron', which was the practical long-wavelength limit for large arrays and also for the lenses which were needed in focusing the sky image onto the detector arrays. 2MASS used three separate arrays, each with 256 x 256 pixels, in the three clear bands at 1.25, 1.65, and 2.17 microns. The telescopes were mounted equatorially, with the advantage that images did not rotate during an exposure. In a novel scanning routine, the telescope made a continuous nodding motion, while the images were held fixed by a contrary movement of the secondary mirror for each exposure time of 1.3 seconds. The whole system of operation was completely automatic and remotely controlled. 2MASS was outstandingly successful. A single night's observation produced around 250,000 point sources, mostly stars, and 2,000 galaxies. The final catalogue contained 300 million point sources and 1 million extended sources, including many star clusters and galaxies which had not been seen before. The majority of the extended sources are galaxies, including some very distant. One of the faintest is also one of the nearest—the Canis Major Dwarf Galaxy, a newly discovered member of our Local Group of galaxies. Already by 2015, several thousand published papers had made use of the 2MASS database.

Building on the success of 2MASS, the UKIRT survey took advantage of the greater collecting area (an aperture of 3.8 metres compared with 1.3 metres), and improved technology to increase the size of the arrays. The optical system of the telescope was modified to provide a larger field of view, with images less than 1 arcsecond in diameter across the whole field. The 16 million pixels of the detector each cover 0.4 arcseconds, and the whole array extends in the focal plane to an area 15 cm across. The array is inside a cryostat cooled to 75K ($-198°$ C). The telescope configuration is a Cassegrain, with a secondary mirror in the centre of the aperture reflecting the beam back along the axis. Unusually, the detector system within its cryostat, and the lenses needed for focusing over the wide field of view, are all located in front of the main mirror. The telescope has in consequence a rather unusual appearance, with a solid tubular housing extending from the centre of the primary mirror almost to the secondary mirror. The lenses needed to provide a wide field and a flat focal area are made of fused silicon, which transmits infrared well up to a wavelength limit just beyond 2 microns.

Even with the multipixel arrays of such a powerful camera, it takes a long time to cover the whole of the sky. The programme of UKIRT surveys which was completed in 2012 included a large-area survey of 4,000 square degrees (which is only one tenth of the whole sky), supplemented by surveys of smaller areas in which longer or repeated observations allowed detection of fainter objects. The extreme example is an ultra-deep survey of an area of only 0.8 square degrees, seeking to detect the faintest possible objects such as extremely distant high-redshift galaxies. The rate of data flowing in these surveys is phenomenal; a single night produces on average more than 2,000 images comprising 100 gigabytes of data. Organizing the reduction of this flow of data into forms accessible to a wide community of potential users is a major task. A catalogue of a billion sources, with positions, intensities in three wavelength bands, and with more detailed spectra and identifications for many sources, is the source for a rapidly growing set of scientific papers; the four-hundredth was published early in 2013.

The advantages of extending observations into the near-infrared band were illustrated dramatically by the discovery at UKIRT in 2011 of the most distant quasar known at the time, recognized by the very large redshift of a major feature of its spectrum to the long wavelength of 1 micron. The spectrum was checked at two large optical telescopes: the William Herschel on La Palma and Gemini North on Hawaii. The redshift was measured at $z = 7.085$, placing the age of the quasar at less than 1 billion years— remarkably young in comparison with the current estimate of 13.8 billion years for the age of the Universe.

VISTA: a Joint Venture

The success of SDSS in surveying the northern sky, and the possibilities opened by 2MASS and UKIRT of extending astronomy into the infrared, inspired the design of VISTA, a 4-metre telescope for the visual and near-infrared spectrum, to be built in the southern hemisphere. UK astronomy had for many years remained independent of the European Southern Observatory (ESO), choosing a partnership with Australia rather than the developing collaboration between most other European countries. However, the declining importance of the AAT, together with ESO's interest in VISTA, provided the opportunity for the UK to join ESO in 2002. VISTA, which was designed and built largely by UK resources, became the entry ticket to ESO, and the telescope is now in operation on an ESO site in Chile, on a subpeak close to Cerro Paranal, where the largest ESO telescopes are

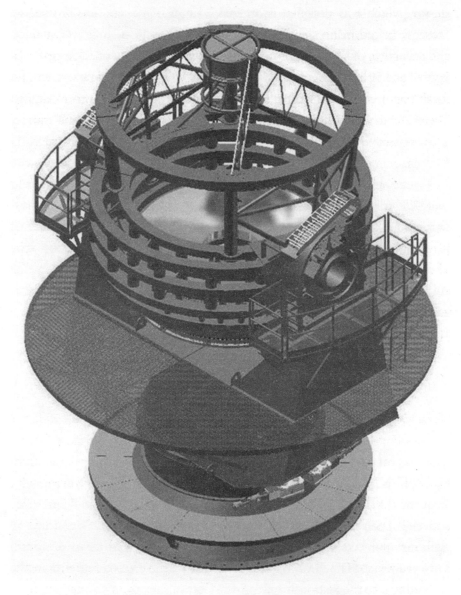

Fig. 26. VISTA, the ESO sky survey telescope. The primary mirror is a deep meniscus of 4-metre diameter, 50 cm deep, and only 17 cm thick.
VISTA

located. From the start, it was designed as an infrared survey instrument, with a wide field of view and a massive multipixel camera.

The international nature of VISTA is illustrated by its 4-metre primary mirror, which was designed in the UK, cast in Germany, and figured in Moscow before being shipped to Chile. It is a meniscus, only 17 cm thick and deeply curved, like a dish 0.5 metre deep—the deepest optical mirror ever figured and polished. Figure 26 shows that the very small focal ratio of f/1 (the focal length equals the diameter) allows the telescope to be very compact. There are 105 individual supports behind and on the edge of the mirror, allowing active control of its shape while observing. The optical system is the conventional Ritchey–Chretien, with a 4-metre hyperboloid primary and 1.24-metre secondary mirrors, and an additional lens extends the field of view to a width of 1.65°. With stellar images only 0.5 arcsecond across, a very large number of pixels are needed to record a full image. No fewer than 67 million pixels are used, comprising sixteen CCD arrays, each of 2,048 x 2,048 pixels. The detector elements are cooled HgCdTe semiconductors, which work efficiently into the near-infrared, and the whole camera system is cooled to 70K. Filters select wavebands between 0.84 and 2.54 microns.

Putting It All Together

The astronomical objects which are uniquely available from these infrared surveys cover the whole field from the Solar System to the most distant galaxies yet observed. A large part of modern astronomical research, however, is based on putting together the surveys at infrared wavelengths with those at optical wavelengths, and in particular the Sloan Digital Sky Survey (SDSS) described in Chapter 3. This classic survey of hundreds of millions of objects in the northern sky used a dedicated 2.5-metre telescope in New Mexico. The SDSS measured intensity in five wavelength intervals, extending into the near-infrared at 893 nm. Handling the data from all these surveys, and from radio surveys, is obviously a task that is only possible using computers. However, computers cannot do the whole job. So many galaxies had never been seen before that there was a serious problem in classifying them and in answering basic questions, such as how many galaxies of different types are there, and whether the spiral galaxies have a preferred direction of winding. Inspecting the individual images turned out to be an enormous task which is done best by human eyesight rather than by machine, and a worldwide network of amateurs was called on to sort out the basic characteristics of galaxy images distributed on the Internet. The project, called Galaxy Zoo, was taken up enthusiastically by hundreds of amateurs, who evidently much enjoyed the chance to be involved in front-line research. One of the home analysts, Hanny Van Arkel, achieved international recognition by spotting a new type of gas cloud close to a distant quasar and probably heated by the black hole at the centre of the quasar. The cloud is now named Hanny's Object. Similar web-based analysis systems have been adopted for other tasks, including the discovery of pulsars hidden within recorded radio observations (Chapter 8).

The original SDSS survey was followed by spectrographic measurements, using the same 2.5-metre telescope, of most of the million galaxies selected from the survey catalogue. To cope with the large number of galaxies, the spectra of hundreds were measured simultaneously, using the fibre-optic system. Repeat observations were used to select variable objects, the spectra of which could be quickly measured. In this way, the survey discovered five hundred new Type Ia supernovae, playing a vital part in the discovery of the acceleration of the expansion of the Universe (Chapter 11).

The Infrared Sky

The infrared sky which is revealed and explored by the new generation of telescopes has been called the Cool Cosmos. The thermal radiation emitted by most objects has a peak intensity at wavelengths depending on temperature; telescopes working at infrared wavebands naturally concentrate on cool objects, with temperatures of order one hundred degrees absolute

(100K) rather than thousands. The division of the infrared wavelength bands into near-, mid-, and far-infrared follows this progression to lower temperatures. The division is in practice defined (although loosely) by the techniques available for the detectors; in particular, the large CCD arrays have so far reached from optical to about 2.5 microns, while at the longer wavelengths bolometers rather than photon detectors have to be used. But in a rough way, the division also applies to the range of astronomical objects which stand out in the three spectral regions.

In the near infrared the sky is easily related to the familiar visible cosmos, with two main differences. First, the dust clouds which obscure the more distant parts of the Milky Way become transparent. This has made possible, for example, the investigation of the centre of the Milky Way, whose visible light is completely lost in dark dusty clouds; the map of individual star orbits within the central arcsecond (see Figure 19 in Chapter 3) was constructed from infrared observations. Second, there are the cool objects which are inconspicuous in visible light; these actually include parts of the same obscuring clouds, which happen to be heated locally by hot stars. Many of these clouds are stellar nurseries, in which gas and molecules are condensing to form new stars. Infrared light also reveals comparatively cool stars, including red giants, and the even colder brown dwarfs, which may be missed entirely in a survey in visible light.

Cold objects such as planets, comets, and asteroids are the prime targets in the mid-infrared—a wavelength region reaching to around 30 microns. This is a region less thoroughly explored; semiconductor detectors do not work well at longer wavelengths, and atmospheric absorption is severe. Cool interstellar dust clouds also are natural targets; they may contain interesting structure which cannot be distinguished at shorter wavelengths. For example, within the dust clouds there may be structure indicating the early stages of formation of stars, or condensations close to existing stars which are forming planetary systems.

The coldest object observed by astronomers, which is also the most intensively studied and the subject of more papers than any other, is the early Universe, seen as the Cosmic Microwave Background (CMB). The

CMB covers the sky with nearly uniform radiation at a temperature of 3K (−270° C). The peak wavelength of radiation from this cold background is around 2 mm (2,000 microns), beyond the far-infrared, at the short-wavelength end of the radio spectrum. Even at this peak wavelength the signal is very weak, as expected from the law relating radiated energy and temperature.

At long-wavelength infrared, where detector systems must be kept cold, it is important also to cool any part of a telescope whose thermal radiation might reach the detector. The best observations are necessarily made from a spacecraft above the terrestrial atmosphere. In such a distant space environment the structure of this spacecraft and its telescope can be cooled naturally to some tens of degrees absolute, while the detectors themselves may need to be further cooled to below 1K. The far-infrared is indeed a difficult and demanding spectral range.

Shorter Wavelengths: the Ultraviolet

If our eyes were sensitive to ultraviolet light, we would see a sky strangely different from the well-known pattern of constellations. Some red stars, like Betelgeuse in Orion, would be nearly lost, while the hot blue stars would dominate the sky. These are stars mostly younger than our Sun, but some are in the final stages of collapse. Away from the Milky Way there would be galaxies standing out through activity associated with black holes at their centres. But for many astronomers the largest reward of ultraviolet astronomy would be to have precise measurements of stellar spectra, extending into a region where many elements show up because of their spectral lines.

Extending the spectrum beyond the short-wavelength end of the visible at 400 nm (0.4 microns) appeals both to astrophysicists and telescope builders. Not only does ultraviolet astronomy promise to open up a new, hot and energetic sky, but the optical techniques of reflecting telescopes and semiconductor detectors are easily extendable to shorter wavelengths.

There is, however, not much space in the spectrum between the visible and an impenetrable barrier to any observations by telescopes at ground level; atmospheric absorption by oxygen and ozone is responsible for a practically complete cutoff at about 320 nm. However, in the narrow spectral band, known as the UVA, between this cutoff and the shortest wavelength visible light there is much to be learned about stars which are hotter than the Sun. These are young stars which are interesting both on their own account and because they are often found to be the brightest objects in the very distant young galaxies now appearing in the infrared surveys. Understanding stars in our own local Milky Way galaxy can help us interpret the processes of formation of the earliest galaxies in the very young Universe.

Astronomy at ultraviolet wavelengths is mostly concerned with spectral lines. Many elements show up in light from hot stars, in which the temperature is high enough to excite energy levels in individual atoms. Ultraviolet radiation from hot stars may ionize gas clouds around these stars and in interstellar space. Interstellar gas also absorbs starlight, so that the spectrum of ultraviolet light from a star may contain information on element abundances both in the star and in interstellar space. Hydrogen is an important example; it has a series of spectral lines in the ultraviolet known as the Lyman-α series. The longest wavelength in this series is 121 nm, well outside the visible spectrum. Hydrogen is the commonest element; before the space age it was particularly frustrating that the characteristic radiation of hydrogen was beyond any possibility of observation from ground-based observatories. Unsurprisingly, the first proposals for overcoming the barrier of the terrestrial atmosphere by using rockets and satellites were for ultraviolet astronomy. The opportunities came immediately after World War II, when sounding rockets for atmospheric research became available, and, soon afterwards, heavier rockets originally designed for missiles could place telescopes in orbit round the Earth. The era of space-based astronomy is the subject of Chapter 5.

5

INTO SPACE

It is surely no coincidence that the human eye is sensitive to exactly the range of wavelengths of sunlight that can penetrate the Earth's atmosphere. Astronomers go to great lengths to overcome the wavelength barriers at both ends of the visible spectrum, building their telescopes on mountaintops to be above the clouds and the lower, wetter part of the atmosphere. This ground-based approach can be very effective for extending the available wavelengths at the red and infrared, long-wavelength end, where the absorption is mainly due to water vapour. There is a real advantage in observing from above the clouds and above a large proportion of atmospheric water vapour. Building telescopes on mountaintops is not much help for ultraviolet light, where the atmospheric barrier is absorption in oxygen and ozone. Even on the highest mountain sites there might still be more than half the atmosphere above the telescope. There is no escaping the necessity for astronomy in ultraviolet light to be based on rockets that can carry telescopes into clear space above the highest levels of the atmosphere, and preferably into orbit. In the infrared, however, it is worthwhile to mount a telescope in an aircraft that can carry a telescope and a team of observers well above mountain observatory sites, flying only a little higher than normal intercontinental airliners. In 1965, Frank Low (1933–2009) was the first to make infrared observations from an airplane, and later he used a 12-inch telescope in a Learjet to make many notable observations of the Solar System planets. A larger joint project of NASA and the German Aerospace Center (DLR) is the Stratospheric Observatory for Infrared Astronomy (SOFIA). This is a successor to the Kuiper Airborne Observatory (KAO), named after Gerard Kuiper

(1905–1973), using a Boeing 747 aircraft which is retired from commercial service and is equipped with a 2.5-metre aperture infrared telescope looking through a huge door in its side. The aircraft can fly for many hours at a height of 45,000 feet (13.7 km). At that height it is possible to observe at wavelengths well beyond the visible red spectrum. The advantage of this airborne observatory over satellite-borne telescopes is accessibility; new detector systems can be tested and developed almost from flight to flight.

The so-called near-infrared extends to wavelengths around 10 microns—more than ten times the wavelength of visible red light. SOFIA can make useful observations beyond the near-infrared, extending to wavelengths of several hundred microns. Furthermore, the new detector technologies discussed in Chapter 4 can be built into useful instruments in a much shorter time than it takes to design, build, and launch a spacecraft. There is certainly room for development: the ideal camera for any wavelength would have a multipixel array of detectors, each of which would measure the energy of each and every arriving photon. The practical situation for infrared wavelengths up to 40 microns is for efficient photon-counting using silicon-based diode arrays; for example, FORCAST, a faint object camera built in Germany for SOFIA, covers the range 5–40 microns. At longer wavelengths the detectors are bolometers, measuring the heating effect of a flow of energy rather than counting photons; FIFI-LS, also from Germany, is a field-imaging far-infrared spectrometer, using a supercooled 16 x 25-pixel array of germanium bolometers.

Although this airborne observatory has effectively opened up the new fields of the near- and mid-infrared, absorption and radiation from the remaining atmosphere above the aircraft remain as serious limitations. There are also some practical difficulties which are not encountered in orbiting spacecraft; the telescope must be guided and kept free from vibration in the moving aircraft, and it must be operated in an isolated compartment where it can be kept cold, while avoiding atmospheric water condensation when the aircraft is on the ground.

Long-wavelength infrared astronomy is concerned with observing cool objects—particularly the planets of our Solar System, which were a particular interest of Kuiper, who proposed the existence of a concentration of small planets (asteroids) beyond the orbit of Neptune, now known as the Kuiper Belt. The KAO is credited with the discovery of the rings of Uranus and the atmosphere of Pluto.

Rocket Science

The idea of rockets that might fly above the Earth's atmosphere may be traced back to before World War II, to individual enthusiasts in three countries: Konstantin Tsiolkovski in Russia, Robert Goddard in the USA, and Hermann Oberth in Germany. The first real success came in 1944 and 1945 with the German military effort to launch V2 rockets, and as it turned out it was a V2 rocket that later made the first astronomical observation from above the atmosphere. This was achieved after the war in the USA, where in 1948 a simple photon-counting system on a V2 rocket detected X-rays from the Sun. This could hardly be called a telescope; it was only possible to detect the origin of the X-rays by observing a periodic change of count rate as the roll of the rocket body brought the Sun into the wide field of view of the photon counter.

The launching of V2 rockets stimulated a campaign of using smaller and more useful rockets for atmospheric and meteorological research. It would be unprofitable to install a conventional telescope in one of these rockets for a ride of only a few minutes above the atmosphere, but some astronomy was achieved nevertheless by using simple detectors for energetic cosmic-ray particles and X-rays. Later, in 1948, Aerobee rockets carried cosmic-ray detectors to a height of over 200 km, well clear of the atmosphere.

In the UK, the first of a series of no fewer than 441 Skylark rockets was launched from Woomera, Australia, in 1957. In the 1960s, the Skylark evolved into a very useful vehicle for ionospheric and astronomical research, especially for X-rays from the Sun.[1] The successful orbiting of Sputnik I, also in 1957, suggested the possibility of using orbiting satellites as a platform for substantial telescopes. It was the prospect of operating in the ultraviolet spectrum that brought astronomers into the space age. Among the first was Robert Wilson[2] (1927–2002) of University College London, who was already an expert in ultraviolet spectroscopy.

The Ultraviolet Sky

A pioneer in the study of the ultraviolet, Robert Wilson, had worked for several years on thermonuclear power generation, in which he investigated the properties of very high energy ionized gas by using ultraviolet spectrometers. His expertise was now to be applied to understanding both the energetic centres of active galaxies, and the much lower energy ionized gas in the space between ourselves and the stars. In the active galactic centres there would be emission lines, and in the interstellar medium there would be absorption lines showing as dips in the continuous spectrum of light from individual stars. The lines would be resonance lines in heavier elements such as barium. A particular interest was the constitution and density of the interstellar medium—not only for its own sake, but to quantify the amount of absorption of light over the whole spectrum. The distances of stars as measured by their brightness depended on correcting for this absorption, which was often large enough to have serious effects on our understanding of stellar astrophysics and star populations.

In 1964, Wilson proposed that a satellite devoted to measuring the ultraviolet brightness and spectra of stars should be launched by the newly formed European Space Research Organization (ESRO). Europe's first astronomical satellite, TD1-A, was launched in 1972 by a Thor-Delta rocket into an orbit at 540 km altitude. Operating at wavelengths between 135 and 255 nm—about half the wavelength of the violet end of the visible spectrum—TD1-A measured the intensity of ultraviolet light from individual stars in three broad bands of the spectrum. The detector was a single photomultiplier, and the spectrum was scanned past it by the rotation of the satellite. The more ambitious project then proposed by Wilson was to measure these spectra in detail, using a satellite-borne spectrometer. This project was initially taken up by ESRO as the Large Astronomy Satellite. In the event, ESRO could not obtain sufficient funds, and the satellite instead became the International Ultraviolet Explorer (IUE), a project primarily of NASA in partnership with the UK and the European Space Agency (ESA, the successor agency to ESRO). IUE was launched in 1978, and remained in continuous operation for eighteen years. An imaginative view of the telescope in orbit is shown in Figure 27.

The main requirement for the IUE was to allow an observer to view a small area of sky, choose the desired target star, and guide the telescope so that the starlight entered a grating spectrograph (see Figure 11 in Chapter 2).

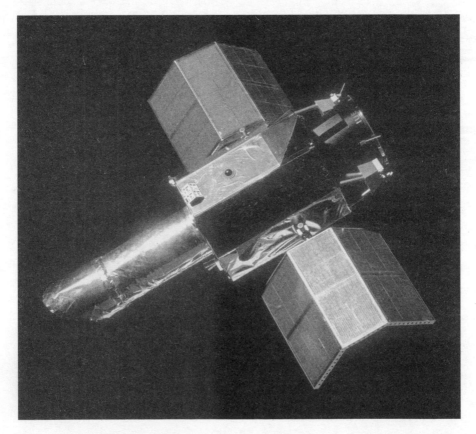

Fig. 27. The International Ultraviolet Explorer (IUE).

The telescope was conventional, with a primary mirror 45 cm in diameter. The stars in a field of about 0.25° across were shown through a vidicon television camera adapted for ultraviolet light. There were two spectrometers, one of which was a reserve in case of failure (though in the event it was the reserve which failed). Long exposures were possible, as the IUE was placed in a perfect environment with no disturbance from sunlight or from the usual temperature variations or vibrations normally encountered in terrestrial observatories. Continuous contact with the ground observing stations was achieved by using a geosynchronous orbit, at a distance varying between 26,000 and 42,000 km. The archive of data from the IUE is still widely in use, and more than 4,000 peer-reviewed papers have been published using IUE data.

Other ultraviolet telescopes were launched in this era, all working in the wavelength range between 100 nm and the short wavelength limit of the visible spectrum at 350 nm. In this ultraviolet range, mirrors and other components were similar to those of optical telescopes, but adapted to the requirements of space research. The IUE used a beryllium mirror, and a rhenium mirror was used in a 75-cm Schmidt taken by Apollo 16 to the Moon in 1972. This Apollo telescope notably observed comet Kohoutek and the terrestrial aurora. Shorter wavelengths needed a different approach, since there was no material available with reasonable reflectivity for use as a primary mirror. Mirrors with up to 50 percent reflectivity have been made for the shortest ultraviolet wavelength range of 17–30 nm, using multilayer optical coatings. These work only for a small range of wavelengths determined by the thickness of the coated layers.

A telescope working in the extreme ultraviolet (EUV) region was installed on the Solar and Heliospheric Observatory (SOHO), launched in 1995. This EUV Imaging Telescope (EIT) was designed to observe the solar corona. With a temperature of more than 1 million degrees, the corona is a bright source of ultraviolet radiation, and particularly of emission lines from highly ionized iron. With such a bright source, a small mirror is large enough. In the EIT the primary mirror was only 12 cm across, and was divided into four sectors, each with its own multilayer coating which reflected selectively one of the iron emission lines or a line from helium. The detector array provided a picture of the whole corona with a resolution of 5 arcseconds (about 1/360 of the angular diameter of the Sun). Apart from its phenomenal success in observing the corona daily for over fourteen years, the EIT observed any bright object which was close to the Sun, and spectacularly discovered more than 2,000 comets.

The Hubble Space Telescope

The original idea of a major telescope operating in space came in 1946 from Lyman Spitzer (1914–1997), astronomer and mountaineer, of Princeton University in the USA. His idea came to fruition in 1990, as the Hubble Space Telescope (Plate 7). Naming this great orbiting observatory after Edwin Hubble was entirely appropriate, for it opened up the Universe beyond our galaxy as he had started to do sixty years earlier. Spitzer himself is commemorated in the Spitzer Space Telescope (SST), launched in 2003.

Launching a large-diameter telescope (the original proposal was for a diameter of 3 metres) into Earth orbit required vehicles which simply did not exist in 1946. The possibilities of multistage rockets had been pointed out by Hermann Oberth (1894–1989) in 1923; notably, he worked for a time with Wernher von Braun (1912–1977), who is credited with the invention of the V2 and the Saturn V rockets, and is considered by NASA to be the father of rocket science. The reality came with the Space Shuttle, which went into service in 1982. This launch vehicle allowed the possibility of manned visits to an orbiting observatory, avoiding the requirement for faultless operation throughout a long prospective lifetime, and allowing changes of instruments in flight (Plate 7). As it turned out, no fewer than five major repair and refurbishment missions were flown to the HST. Without these, the lifetime and success of the telescope would have been severely limited.

The primary mirror of the HST is made of low-expansion coefficient glass, with thin back and front surfaces separated by a honeycomb lattice.

No active adjustment system was needed for operating in the practically zero gravitational field and constant temperature in orbit; only a small focusing adjustment was incorporated. The mirror diameter is 2.4 metres, and it is perhaps no coincidence that this is the same diameter as the mirrors used in the series of spy satellites known as KH-11, of which the first was launched in 1976. The focal length of the HST mirror is 57.6 metres, accommodated by reflecting the light to and fro as in a Cassegrain, and long enough to provide a large-scale image for the instruments at the focal plane. The mirror therefore needed only a shallow curvature, but it had nevertheless to be ground and polished to an accuracy related to the shortest wavelength of light to be detected; that is, the near ultraviolet. The mirror surface apparently met a very tight specification, within 10 nm. Ready for launch, it was given a thin reflecting coat of aluminium protected by magnesium fluoride. It was the most precise mirror ever made, but unfortunately it was the wrong shape—a disaster which was discovered only after the telescope was already in orbit.

During the grinding and polishing of a large mirror, the shape of the surface is measured by constructing a test piece in the form of a smaller mirror or lens with a precise shape which is opposite to that required in the large mirror. Light passing through both the mirror and the test piece should then produce a perfectly plane wavefront. Combining this with another plane wavefront produces an interference pattern which shows every detail of the mirror shape. The specification for the HST mirror was, as usual, for a Ritchey–Chretien hyperbolic shape, to be met with an accuracy approaching 10 nm. The test piece was very accurately constructed, but unfortunately was placed at the wrong distance from the mirror. The result was that the outer part of the mirror was too flat; the perimeter was in error by no less than 2,200 nm. The effect on a star image was obvious; the first images showed a sharp centre surrounded by a large diffuse halo. There seemed to be no remedy.

By the time of launch in 1990 the HST had cost more than 2.5 billion dollars, part of which had been contributed by the European Space Agency. A public-relations triumph for astronomy had apparently turned into a disaster. There was, however, a remedy, given that there would be access from visiting astronauts. The remedy was related to the original test procedure. If a thin lens could be inserted into the convergent light beam from the mirror, the shape of the wavefront could be corrected. Now the wisdom of allowing for manned visits to HST while in orbit came to the rescue. At the first visit of the Space Shuttle, in 1993, astronauts fixed in place the necessary optical corrector, resulting in perfect images and a complete restoration of confidence both for the observers and the general public. The only loss was one of the four original instruments, a high speed photometer, which had to be removed to make way for the optical corrector (but not before I had used it for observations of the Crab pulsar). The corrector itself was removed in a Space Shuttle visit in 2002, when new instruments were installed incorporating their own optical correctors.

The HST is the only space telescope that has been accessible for servicing in orbit. Most others are in deeper space orbits, well outside the range of the Space Shuttle, and the expense of manned visits has been transferred to more rigorous project management and repeated testing before launch. Considering the early date of 1978, when the funding of the HST was approved by Congress and the overall design effectively frozen, the performance has been phenomenal. From the start, a prime objective has been to inform, educate, and inspire the public; HST pictures in full colour of all manner of galaxies and spectacular nebulae within our galaxy are familiar to every newspaper reader worldwide. Scientifically, the HST has transformed observational astronomy, producing on average more than two scientific publications per day. The outstanding observations must be those penetrating furthest into the depths and history of the Universe, in a series concentrating on areas of sky known as the Hubble Deep Field, beginning in 1995–1996.

Observing time on the HST is allocated by a Programming Committee. The Director of the Space Telescope Science Institute is, however, allowed a proportion of the observing time for programmes of his own choosing. In a large part of the years 2003–2004 the Director chose to devote the whole of his time to a single long exposure of a small piece of dark sky, to observe

Plate 1. The first reflecting telescope, built by Isaac Newton in 1668.

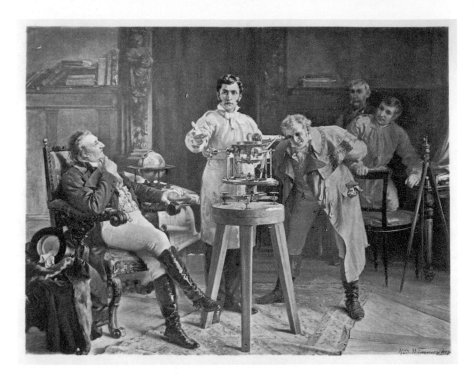

Plate 2. Joseph von Fraunhofer demonstrating the spectroscope. Photogravure from a painting by Richard Wimmer.

Plate 3. The four 8-metre telescopes of the Very Large Telescope (VLT) at Cerro Paranal, Chile.

Plate 4. The Mauna Kea observatories, Hawaii. On the foreground ridge, left to right: the UK Infrared Telescope, the University of Hawaii 2.2-metre telescope, the Gemini North 8-metre (with open dome), and the Canada–France–Hawaii 3.6-metre telescopes. At right: the twin Keck 10-metre telescopes, with the Subaru 8-metre telescope behind. The 15-metre James Clerk Maxwell sub-millimetre telescope (with open dome) is in the valley between.

Plate 5. Images of Uranus obtained with the Keck telescopes on Mauna Kea, using adaptive optics. The detail of the ring and the surface features could not be seen in ordinary ground-based observations.

Plate 6. The Sloan Digital Sky Survey telescope. To ensure the best seeing conditions, the whole telescope enclosure slides to one side, leaving the telescope exposed to undisturbed air.

Plate 7. The Hubble Space Telescope at its release from the Space Shuttle *Discovery* in 1990. The front cover is still closed.

Plate 8. The Hubble Ultra Deep Field. This small dark patch of sky, only 2.3 × 2 arcminutes, contains many thousands of galaxies, some of which are observed as they were in the first third of the history of the Universe.

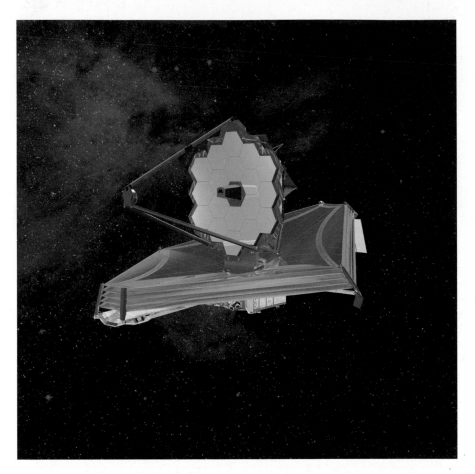

Plate 9. Artist's impression of the James Webb Space Telescope, showing the segmented mirror and the large sunshield. Launch is planned for 2018.

Plate 10. The spectacular spiral galaxy M31 in Andromeda, imaged in its glorious infrared colours by Spitzer. The observations were carried out with the Multiband Imaging Photometer for Spitzer (MIPS), which observes at 24, 70, and 160 microns. Two huge waves of star formation encircle its central nucleus, producing beautiful spiral arms. Each one shines brightly, with its dust being warmed by the young stars.

Plate 11. The Eta Carina Nebula imaged by Herschel in three infrared wavelengths, 70, 160, and 250 microns, shown as blue, green, and red, covering approximately 2.3° × 2.3°. The image shows the effects of massive star formation; powerful stellar winds and radiation have carved pillars and bubbles in dense clouds of gas and dust.

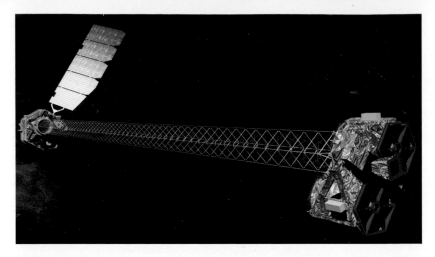

Plate 12. The Nuclear Spectroscopic Telescope Array (NuSTAR). The 10-metre long lattice mast which connects the X-ray optics (right) with the detector unit (left) was folded into a canister 1 metre long at launch. The telescope is in a near-Earth orbit. The detector is a 32 × 32-pixel CCD, in which each pixel acts as a spectrometer.

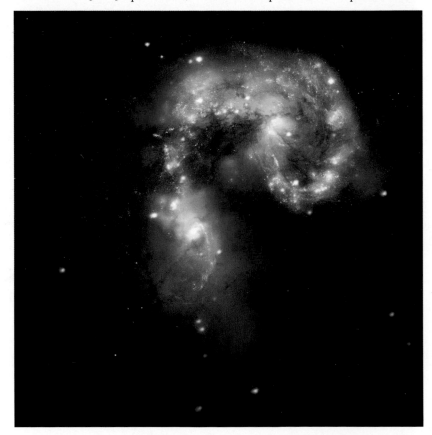

Plate 13. A beautiful new image, created by NASA's Great Observatories, of two colliding galaxies. The Antennae galaxies, located about 62 million light-years from Earth, are shown in this composite image from the Chandra X-ray Observatory (blue), the Hubble Space Telescope (gold and brown), and the Spitzer Space Telescope (red). The Antennae galaxies take their name from the long antenna-like arms, seen in wide-angle views of the system.

the faintest possible images of the most distant galaxies. The observing time totalled more than 1 million seconds, extending over four hundred orbits (the orbital period is 96 minutes). The observations in this Hubble Ultra Deep Field survey were repeated in three colours, with an extension into the near-infrared at a wavelength of 850 nm. The area of sky was tiny: only 3 arcminutes across, about one tenth of the angular diameter of the Moon. Within this small area, no fewer than 10,000 objects were found, most of which could be distinguished as individual galaxies (Plate 8). Adding the near-infrared observations in 2012, using new instrumentation installed in the last servicing mission, brought the total to 15,000 galaxies.

When Edwin Hubble found that galaxies are, on average, receding from us at a rate which increases with distance, he was looking at only a very small, local part of the Universe. A receding velocity has the effect of shifting the spectrum towards longer wavelengths; this is the redshift, z, measured as a ratio of the wavelength shift to the emitted wavelength. The galaxies which Hubble and his followers were dealing with had redshifts much smaller than 0.1; the HST was opening up an unexplored region in which galaxies with redshifts of $z = 6$ or 7 were appearing. These enormous redshifts moved the wavelengths of a typical spectrum so far that most of the light was seen as infrared, and the galaxies could hardly be seen in visible light. In fact, the redshifts could be estimated from the ratio of light detected in infrared as compared with the three colours in visible light. A prominent feature in the spectrum of many galaxies is the 'Lyman break'—a characteristic step in the spectrum due to absorption by neutral hydrogen in the ultraviolet at 91 nm. Large redshifts bring the Lyman break into the visible or even the infrared spectrum, and the observed wavelength of the step in the spectrum is a good indication of the redshift of a galaxy.

The highest-redshift galaxies, at $z = 6$ or 7, are those that are being observed at a very early stage in the development of the Universe, around only 1 or 2 billion years since the Big Bang 13.8 billion years ago. Comparing the distant and near galaxies in the Hubble Deep Field provides clear evidence of evolution. Two improvements to Hubble's law of expansion could now be made: the ratio of rate of expansion to distance (the Hubble

Constant) was measured much more accurately, and, more fundamentally, we could investigate whether Hubble's law itself continued to apply at such large distances. Cosmology had become accessible to observation rather than remaining a subject for theoretical speculation (more about this in Chapter 11).

The James Webb Space Telescope

The emphasis on infrared pointed the way to the HST's successor: the James Webb Space Telescope (Plate 9). Originally planned in 1996 as the Next Generation Space Telescope, it was renamed for James Webb (1906–1992), who was an outstanding administrator of NASA from 1961 to 1968. The JWST project is led by NASA, with Canada and the European Space Agency as partners, with many other participating countries. Like the HST, it has suffered severe delays, mainly due to a runaway budget: the US Congress temporarily halted the project in 2011, when $3 billion had already been spent, and allowed it to continue with a budget capped at $8 billion. The launch, on an ESO Ariane launcher, is set for 2018.

As we saw in Chapter 4, the detectors on infrared telescopes are liable to pick up thermal radiation from the telescope itself, and particularly from the mirror. The JWST will have a very large and very cold mirror, and will be in an orbit far away from the Earth's infrared radiation. The specifications for the mirror and the detectors are among the most demanding requirements for any spacecraft, and, unlike the HST, they must be met without any recourse to servicing missions. The mirror diameter is 6.5 metres, which cannot be accommodated as a single piece in any launch vehicle. It has to be made of eighteen hexagonal panels, made of polished beryllium, to be unfolded when the telescope is in orbit, and adjusted to make a single mirror. The whole surface of the assembled mirror must be accurate to a small fraction of the wavelength of red light (not quite as accurate as the HST mirror, which was designed for ultraviolet light). No mechanism could do this without allowing further adjustment after deployment. To make this adjustment, each panel will be moveable by a tiny motor. The plan is to examine in detail the image of a bright star and deduce the position of each panel from the shape of the image. This is similar to the task of correcting the defect in the HST, but is more complex, since in the JWST the deviations from a perfect hyperboloid could be randomly distributed rather than symmetrical. The measurement process is known as wavefront analysis. A single image is insufficient, so several image shapes must be measured while refocusing the telescope. Once the mirror is correctly aligned, the advantage of practically zero gravity removes any necessity to readjust the mirror as the telescope is moved across the sky.

The JWST will be placed in a special orbit at a distance of 1.5 million km from Earth, on the opposite side of the Sun. This is a stable position in which the combined gravitational pull of the Sun and Earth keeps a spacecraft orbiting in the same relationship, moving round the Sun once every year. In 1808, the mathematician Joseph-Louis Lagrange showed that there are five such stable positions (see Figure 28). The first three were discovered by Leonhard Euler in 1765; they lie on the Earth–Sun line. L4 and

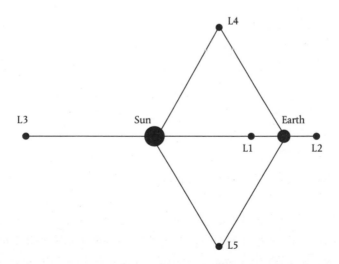

Fig. 28. The five Lagrangian points in the Sun–Earth system. The diagram is not to scale: the points L1 and L2 are both 1.5 million km from Earth, while the distances from the Sun to Earth and to L3, L4, and L5 are close to 150 million km.

L5 each form an equilateral triangle with the Earth and the Sun. The outer position, L2, is particularly useful for spacecraft, as the Earth and the Sun are together in the sky, and leave more of the sky unobscured.

The JWST will actually move in a small orbit around the outer position, designated L2. The telescope will avoid pointing anywhere near the Sun and Earth, and a huge aluminized plastic sunshield will keep it cold. The mirror temperature will be only 40K, which is ideal for sensitive infrared detectors.

When the JWST is finally in operation it will be one of NASA's four Great Space Observatories, joining the HST and the X-ray and gamma-ray telescopes, the subject of later chapters. But meanwhile, infrared astronomy has not been neglected: a succession of space missions have been launched which have surveyed the infrared sky, with less sensitivity than the JWST, but nevertheless with spectacular success.

IRAS, ISO, Spitzer, Akari, and WISE

The first infrared observatories were placed in near-Earth orbits, chosen to allow a scan of the whole sky without pointing too near the Sun, Earth, or Moon. In a polar orbit, the plane of the orbit can be perpendicular to the direction of the Sun, allowing the telescope to point outwards and scan a great circle in the sky. Throughout the year, as the Earth moves round the Sun, this circle moves to cover the whole sky. The Infrared Astronomical Satellite (IRAS), launched by the European Space Agency (ESA) in 1982, Akari (Japanese for 'light'), launched by Japan in 2006, and the Widefield Infrared Space Explorer (WISE), launched by NASA in 2009, were in nearly circular orbits between 500 and 1,000 km altitude. The Infrared Space Observatory (ISO), launched by ESA in 1996, was in a highly elliptical orbit, with the furthest point (apogee) at 70,600 km and the nearest (perigee) at 1,000 km, producing a 24-hour orbit.

All these had comparatively small telescope apertures, between 40 cm and 60 cm diameter—adequate for a preliminary survey. Infrared telescopes have to be kept cold, and these telescopes were small enough for the whole telescope and its detector system to be enclosed in a single cryostat at a temperature of a few degrees absolute. Cooling required a tank of liquid helium (or solid hydrogen for WISE), providing a limited life of only one or two years until the coolant had evaporated.

IRAS was the first to survey the whole sky, producing a catalogue of 350,000 sources which is still widely used; many were newly discovered and not yet identified in other observations. Apart from the cold bodies such as asteroids in the Solar System and dust rings round stars, IRAS discovered intense radiation from some very active galaxies known as starburst galaxies. These are obscured from normal view by dust clouds, and we can see them only by observing their infrared light.

Similar technologies were used across all of this series of comparatively small infrared telescopes. Two notable innovations were introduced in Akari—in the cooling system, and the material for the mirror. No evaporating coolant was used; instead, there was a continuously operating closed cycle refrigerator. The mirror used a new material: silicon carbide (SiC)—a ceramic material which is very stable mechanically and thermally, and which can be polished like glass. SiC was later used to make the mirror for Herschel, a much larger telescope which we will examine later. All these explorations of the infrared sky showed how rich a field was opening up for astronomy. The next stage required a much improved rate of data collection and transmission, and eventually an orbit taking the telescope further from the disturbing thermal radiation from Earth. The improved data rate was the basis of WISE, while Spitzer was the first to explore the possibilities of deeper orbits.

WISE, launched in 2009, built on the experience of the earlier missions, set out to construct a much larger catalogue of infrared sources over the sky The key advance was the use of new types of detector arrays which had been developed for the JWST, increasing the total number of pixels to 1 million. These were tuned to four wavelength bands. In the near infrared, at 3.4 and 4.6 microns, the familiar HgCdTe detectors could be used, while for the longer wavelengths, at 12 and 22 microns, a new silicon arsenide (SiAs) technology was used. With these more sensitive multi-element detectors, and the comparatively wide field of view (47 arcminutes), the overall sensitivity and rate of observing were orders-of-magnitude improvements on the earlier missions. Even when the solid hydrogen in the cooling system had all evaporated, there was sufficient sensitivity to make it worthwhile to continue operations in a dedicated search for asteroids and other objects in the Solar System. WISE was reactivated in 2012 for this special purpose, which included a search for objects with orbits that might bring them close to Earth.

Scanning the whole sky with unprecedented sensitivity, and with an angular resolution of 6 arcseconds, WISE has provided a catalogue of no fewer than 750 million objects. Comparing intensities in the four wavebands allows these to be sorted into cool Solar System objects, cool stars like the brown dwarfs, and distant, very energetic galaxies. Some of these galaxies have a large population of young and very bright stars, and are known as starburst galaxies. They are observable by telescopes such as WISE at redshifts greater than $z = 1$, and are sufficiently distant to be interesting to cosmologists exploring the formation of galaxies in the early Universe.

The last of the series of comparatively small infrared space telescopes was Spitzer, named after Lyman Spitzer. This telescope evolved from a concept for a large spacecraft intended for launch by the Space Shuttle, but the Challenger disaster made that impossible. Spitzer was launched in 2003 into an orbit which would take it far from Earth, so that most of the necessary cooling was achieved naturally by its remoteness from Earth. This orbit was a Sun-centred orbit, like that of Earth; in fact, the spacecraft was trailing Earth in its orbit round the Sun. The result was a natural cooling to around 30K, and a tank of liquid helium brought the temperature of the primary mirror and the detectors down to a few degrees absolute. The 85-cm mirror was made of beryllium. Spitzer extended the useful wavelength range to 180 microns, and produced some spectacular pictures of galaxies, including the Andromeda Nebula (Plate 10).

The spectra of distant galaxies obtained by Spitzer and the Hubble Space Telescope were the first contributions to a project to combine observations of distant galaxies: the Great Observatories Origin Deep Survey (GOODS). This massive project now incorporates observations by the Herschel spacecraft in the far infrared and the Chandra and XMM Newton X-ray telescopes (Chapter 6).

The WISE catalogue will be the definitive source catalogue for objects that will be the targets for telescopes with higher angular resolution, which is achievable only by using telescopes with larger-diameter mirrors. A large aperture is the main purpose of the JWST, but another mission has already made a step in this direction. This is Herschel, named appropriately after the discoverer of infrared radiation (see Chapter 4). (His name is also attached to a major Earth-bound telescope: the 4.2-metre William Herschel Telescope on La Palma, in the Canary Islands).

Herschel in Orbit

Building on the several innovations and the success of the group of small infrared telescopes, in 1982 the European Space Agency decided to construct a major space telescope for the far infrared. This was proposed as the Far Infrared Space Telescope (FIRST), which was to have a large cooled mirror and operate in a cold environment far from Earth, later specified as the Lagrangian point L2. In 2000, the telescope was renamed Herschel (Figure 29). From the start, it was an ambitious project, requiring the large Ariane launch vehicle to carry a mirror 3.5 metres in diameter to a distance of 1.5 million km from Earth. Furthermore, the same launch was to include another ambitious project, the cosmic background telescope Planck (Chapter 11), which was also to operate at L2. It says much for the reliability of space technology that ESA could even contemplate launching two such large and expensive spacecraft on a single vehicle, but there were advantages to be gained from a common destination and a unified design of spacecraft structure and telemetry systems. The launch was executed perfectly, and both projects were an outstanding success.

Fig. 29. The Herschel infrared telescope. The thin 3.5-metre mirror is protected from the Sun by a sunshield which is covered by solar cells.

The 3.5-metre mirror was the largest that could be accommodated in one piece in the Ariane launch vehicle. Conventional materials—even beryllium, which is used in the segments of the JWST—would be too heavy in such a large mirror. The new ceramic material silicon carbide, pioneered in Akari, could be moulded to be only 1 cm thick and stiffened with ribs on the back. But the largest piece that could be made was only 1 metre in diameter. The solution was to develop a brazing technique to join twelve pieces to form the lightest large mirror ever constructed, weighing less than 300 kg. The history of this mirror was typical of the international nature of the European Space Agency; the silicon carbide was cast in France, ground to shape and polished in Finland, and given its reflecting coat in Spain. At its remote orbit near L2, the mirror was cold enough for operation at mid-infrared wavelengths, although the detectors had to be cooled further by liquid helium.

Herschel operated at the long wavelengths of the far infrared, where there is no possibility for the usual CCD detector arrays to function, since the energies of individual photons are too low. Herschel's detectors reached a wavelength of 672 microns (almost 1 mm); this is regarded as lying in the sub-millimetre band, which we will encounter again as the short wavelength end of the radio astronomy wavelength regime. From 194 to 672 microns the detectors are similar to radio receivers, using a system which can measure a spectrum with considerable accuracy. At the shorter end, from 55 to 210 microns, the detectors were bolometers, which measure the flow of energy rather than the arrival of individual photons. The bolometers were tiny superconducting metal films, which become slightly resistive when radiated. These worked only at very low temperature, below 1K. All the detectors were cooled by liquid helium, which slowly boiled away, giving the telescope a maximum working life of just over three years. Herschel operated from 2009 to 2013, when it was shut down and allowed to drift away from L2 to a solar orbit, where it will remain for the foreseeable future.

The results of three years of concentrated operation of Herschel will be analysed for many years to come. Herschel's results lie in many fields, from the structure and composition of galaxies in the early Universe, to the process of star formation, and to the surfaces and atmospheres of planets. Plate 11 shows the Eta Carina Nebula imaged by Herschel in three infrared wavelengths—70, 160, and 250 microns, shown as blue, green, and red—covering approximately 2.3° × 2.3°. The image shows the effects of massive star formation; powerful stellar winds and radiation have carved pillars and bubbles in dense clouds of gas and dust.

Hipparcos and Gaia

Spacecraft provide a far better platform for precise measurements than the traditional terrestrial observatories; there is no atmosphere to refract and distort star images, and no disturbance from changes in temperature or other environmental factors—including the effect of the observers themselves. Stability and consistency during continuous observations over periods of years have transformed traditional measurements of stellar brightness and positions.

The two major space projects Hipparcos and Gaia are part of the long history of astrometry—the science of determining, by precise measurement, the positions, motions, and distances of stars in our galaxy and beyond. Hipparcos was the first. The name was an acronym (High Precision Position and Parallax Collecting Satellite), but the choice of name celebrated the remarkable early achievements of Hipparchus of Nicea, a pioneer of celestial geometry who in the second century BC discovered a fundamental parameter in positional astronomy: the precession of the equinoxes.

Before the space age, traditional astrometry relied on the rotation of the Earth to scan the sky, measuring the time at which individual stars crossed the north–south meridian line. The Airy Transit Circle at Greenwich (Chapter 2), with an accuracy well below 1 arcsecond, was for a whole century the best instrument that could be built for astrometry. There would be great benefits resulting from improving accuracy by a large factor, but this was impossible for an Earth-bound telescope. There were several limiting factors:

the atmosphere distorted and refracted the star images, the structure of the telescope was sensitive to temperature changes, and observing was slow and tedious. All these were overcome by the move into space.

Astrometry from a spacecraft is entirely detached from the rotation of the Earth. All the limitations of the transit telescopes are removed, with no atmosphere and no temperature changes to overcome. No moving parts are needed; scanning the sky is achieved by a slow rotation of the spacecraft, and a slow movement (precession) of the rotation axis. Furthermore, the observer is replaced by a detector system which can deal with many objects simultaneously, constructing a huge catalogue of accurate positions in a few weeks. Repeated measurements of the same area show movements of the stars, either from actual star velocities or the apparent changes known as parallax, due to the changing view of the sky as seen from different parts of the Earth's orbit round the Sun. Parallax is vitally important in determining the distances of stars, but it cannot be measured precisely enough from the ground. The 'pp' in Hipparcos indicates that precise parallax was a major objective of this telescope.

Hipparcos operated entirely independently of the traditional Earth-based astrometric system. Instead of using the rotation of the Earth and constructing a coordinate system based on the axis of the Earth, Hipparcos was spinning slowly on its own axis, and constructed its own system of coordinates. Like the transit telescope, it observed the transit of stars across the focal plane of a telescope. Unlike the transit telescope, it set up its new coordinate system by observing two widely separated areas of sky simultaneously, eventually establishing a coherent system of star positions which covered the whole sky. The two areas of sky, 58° apart, were observed by a single small telescope of 29 cm aperture, with a double beam-combining mirror. The spacecraft rotated slowly, at 11.25 revolutions per day, scanning a great circle in the sky; a slow precession of the spin axis provided complete coverage of the sky 6.4 times every year. The spacecraft was intended for a distant geostationary orbit—the type of orbit used by the communications satellites which appear to hover over the Earth and provide constant television services. Unfortunately there was a near disaster in the launch sequence, and Hipparcos ended up in a highly elliptical orbit. Although this complicated the analysis, Hipparcos worked well from 1989 to 1993, producing a catalogue of the positions and parallaxes of 118,000 stars to unprecedented accuracy. It also demonstrated the potential for this new technique to be expanded to far greater accuracy and to cover many more stars. The result was the astrometric space telescope known as Gaia, designed and built by the European Space Agency, and in full operation from mid-2014.

The name Gaia was originally an acronym, with the letter I denoting Interferometer, which was part of the original proposal. The interferometer idea was dropped, but the project kept its attractive name (Gaia was the primal Greek Mother Goddess). Following the basic idea of Hipparcos, the Gaia spacecraft was designed with two beams scanning sequentially along a great circle. In this new design, two separate telescopes are set apart at the wide angle of 106°, and their separate images are projected onto a single focal plane and detected together by a huge array of CCD detectors. This array of solid-state detectors comprises 106 CCDs, each with 9 million pixels, producing a total of nearly 1 billion pixels. Part of this array records images from the separate telescopes, with photometric information, and another part records spectral line measurements to produce precise line-of-sight velocities. The combined image from the two telescopes drifts across the CCDs as the spacecraft rotates, and the readout system keeps pace exactly with this movement so that the images accumulate and intensify as they move. They are digitized and stored as they reach the edge of the detector system. There are many innovative details in the spacecraft and the telescopes which improve on Hipparcos; for example, the mirrors are made of silicon carbide and are much larger, at 1.45 x 0.5 metres. A very important improvement is the location; like the recent infrared telescopes, Gaia is in a distant orbit around Lagrangian L2, free from any terrestrial disturbances and at a practically constant temperature.

Gaia is expected to locate more than a billion stars to unprecedented accuracy. For most of these, the parallactic motion should be observable; this is the apparent annual cyclic change in position as the Earth (and Gaia with it) moves in orbit round the Sun. Parallax measures distance, so that distances of stars as far away as the centre of the Milky Way can be measured. The actual motion of many of these stars will also be observed, both as a lateral movement, and as a Doppler shift of the spectroscopic line radiation, which produces the line-of-sight velocity. The measured parallaxes will produce distances of about 20 million stars to 1 percent accuracy, and 200 million to 10 percent accuracy, providing a three-dimensional map of the galaxy at a level of detail previously thought to be unattainable. Furthermore, the spectral types of all the stars will be measured by an on-board low-resolution spectrometer. Other objects will naturally be included in the survey, such as asteroids in the Solar System, and some of the planets now known to be orbiting stars other than our Sun. Gaia is in fact a complete observatory, incorporating all the techniques of astrometry, photometry, and spectroscopy, which have occupied the lifetimes of countless astronomers for the last two centuries.

Handling the flow of data from Gaia, and reducing it to a useable form, is a major task for several teams of astronomers. The total quantity of data over the five-year mission will be more than 200 terabytes, transmitted to Earth at the rate of 5 megabits per second.[3] A large ground-station radio telescope is needed to handle such a large data rate from such a large distance, and two 35-metre dishes in Spain and Australia are spending eight hours every day on this task. Gaia's scientific programme began in July 2014, when it produced its first discovery of a supernova, Gaia14aaa, in a distant galaxy.

The independent system of star positions established by Hipparcos and Gaia has eventually to be tied in with the traditional coordinate system. This requires an interesting synthesis of several branches of astronomy, including a vital contribution from radio interferometers (Chapter 10).

Hunting for Planets

Kepler, named after Johannes Kepler and launched by NASA in 2009, has the traditional task of searching for variations in the brightness of stars, but on a scale and with an accuracy unimaginable for ground-based telescopes. This spacecraft was devoted to the search for exoplanets orbiting bright stars. Covering a single 100-square-degree area of sky, Kepler monitored the brightness of each star for many hours or days, looking for the slight reduction in brightness as any planet crossed the disc of the star. Large numbers are involved: 145,000 stars were monitored with very high photometric accuracy.

With a 1.4-metre mirror, Kepler is not a large telescope, but it has a massive 95-megapixel CCD detector and measures stellar intensities to an accuracy of thirty parts in a million. This accuracy allows it to detect small diminutions in brightness as a planet crosses the disc of its parent star. In four years of full operation (2009–2013), Kepler discovered more than 1,000 exoplanets.

Kepler is another example of the application of very large computer power. At the time of launch, the 95-megapixel camera was the largest in orbit. As the requirement was to detect changes in apparent brightness, there was no need to transmit all basic photometric data to an Earth station; instead, a large on-board computer did most of the analysis. The spacecraft was effectively an autonomous observatory. It was placed in an orbit round the Sun similar to the Earth's orbit, but well away from Earth itself.

Space astronomy has developed through many exploratory telescopes to a series of major observatories. Some, like Gaia and Kepler, are in space because of the cold and stable environment, isolated from atmospheric and thermal disturbance. Others, such as the infrared telescopes Herschel and JWST described in this chapter, operate in parts of the electromagnetic spectrum for which the terrestrial atmosphere is a complete barrier. We next turn to X-ray telescopes, which also have to function above the atmosphere.

6

X-RAYS FROM SPACE

The idea that X-rays can penetrate through opaque substances such as human flesh might suggest that they ought to penetrate through the atmosphere, allowing us to look at radiation from very hot or energetic objects in the cosmos. Unfortunately, the atmosphere is a practically impenetrable barrier to all high-energy radiation, from the ultraviolet end of the visible spectrum right through the shorter wavelengths of X-rays and gamma rays (which is unfortunate for astronomy, but without this barrier there would be no human life and no astronomers!). Although the existence of gamma rays in space was known from the showers of high-energy particles created when individual gamma-ray photons hit the atmosphere, and although it was suspected that the Sun might be a prolific source of X-rays, it was not until the beginning of the space age in the mid-twentieth century that any direct observations could be made.

Telescopes for mapping the sources of X-rays in the sky are very different from those in visible radiation. X-rays do indeed penetrate most materials, but are also absorbed as they do so; the lenses and mirrors of our conventional ground-based or orbiting telescopes are useless. The trouble is that X-ray photons are very energetic. Their high energy means that they can easily be detected, using several types of detector such as photographic plates or CCDs, but they cannot be focused by lenses or conventional mirrors. There are two ways round this problem—one using the geometry of pinhole cameras, and the other using a very unusual type of mirror which does work for low-energy X-rays.

As in any part of the electromagnetic spectrum, X-rays may be characterized either by wavelength or by the energy of their photons (Figure 30).

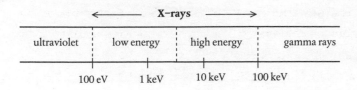

Fig. 30. The X-ray spectrum.

The wavelengths are very short—comparable with the spacings between atoms in a crystal structure. X-ray wavelengths are measured in nm (10^{-9} metre), a unit one thousandth of the micron (10^{-6} metre), which is commonly used as a unit for visible and infrared light. The energies of X-ray photons are correspondingly higher. The X-ray machines used in hospitals produce X-rays by accelerating electrons in strong electric fields, and the photon energies of these X-rays are measured in terms of the voltage accelerating the electrons, typically around 1,000 volts (1 kilovolt). The photon energy of an X-ray is usually measured in terms of the equivalent energy of an electron; the typical energy would then be quoted in electron volts (eV) as 1 keV. The X-ray range is very wide, from around 100 eV to 1000 keV, or 1 MeV. Higher energies belong to the gamma-ray region, which is the subject of Chapter 7. The boundary is not well defined. Perhaps the best distinction is in terms of technique; gamma-ray telescopes cannot produce focused images, whereas X-ray telescopes can do so, as we will now see.

The First Surprises

The first discoveries of cosmic X-rays, using rockets, were made not with cameras but with the simplest of detector systems, with almost no directional discrimination. A detector acting rather like a conventional light meter, first mounted on a V2 rocket in New Mexico in 1949, and later on Aerobee rockets, scanned the sky as the rocket body rotated. Any source of X-rays within a wide angle, around 100°, would be recorded, and there would be no indication of its size. This was good enough for showing that the Sun is a powerful X-ray emitter. There seemed to be other sources, however, which could not be isolated using such a wide-angle detector. In 1962, an attempt to pick out X-rays from the Moon failed, but discovered instead a totally unexpected and very bright X-ray source somewhere towards the centre of the Milky Way. This discovery, by Riccardo Giacconi (b.1931), eventually led him through a series of discoveries to a Nobel Prize in 2002. What could this strange object be? The location was in the constellation of Scorpius. This source, now known as Sco X-1, turned out to be the prototype of an entirely new type of star. But in 1962 there could only be speculation: was it a huge cloud of hot gas, or a tiny source radiating more powerfully than anyone had imagined?

The lead of the Soviet Union in space science, with the Sputnik satellites in 1957 and Yuri Gagarin's orbital flight in 1961, had stimulated the creation of NASA, the National Aeronautics and Space Administration, in the USA. NASA rapidly became the leader in this exciting new field of astronomy, using techniques developed by AS&E, the American Science and Engineering group associated with the Massachusetts Institute of Technology. The Naval Research Laboratory (NRL) was already developing techniques for space research. At NRL a group led by Herbert Friedmann (1916–2000) confirmed the existence of Sco X-1, and discovered another X-ray source, which they tentatively identified with the Crab Nebula. This nebula is an expanding cloud of hot gas, the visible remains of a supernova explosion. It was already known to be a powerful radio emitter, but with the crude detection systems then available, X-ray identification was uncertain. Again, some idea of angular size was needed to be certain that another type of X-ray source had been discovered. A surprising opportunity arose, provided not by new instrumentation but by the Moon.

It happened that in 1964 the Moon's monthly circuit of the sky took it across the Crab Nebula. The window of opportunity was short: if there was to be an observation of the way in which the edge of the Moon cut across the X-ray source, a rocket had to be in place within a few minutes of the predicted time. A perfect launch by the Naval Research Laboratory group achieved this, recording a slow decline of intensity as the Moon moved across the source. The identification of the X-ray source with the Crab Nebula was secure.

Unfortunately, the Moon was not much use for measuring the size of the unidentified X-ray source Sco X-1, as its location was so inaccurately known that no lunar occultation could be predicted. However, AS&E introduced the idea of placing a mask in front of the detector which would show whether a source like Sco X-1 was small or large. The mask consisted of two identical wire grids, spaced apart along the line of sight. X-rays travel in straight lines and do not penetrate metal wires, so that X-rays from a small source form a shadow pattern on the second wire grid. As the rocket carrying the detector rotated, the shadow pattern moved across the second grid, which alternately blocked and let through the X-rays. This modulation was the sign of a source with a small angular diameter. It turned out that Sco X-1 was indeed a remarkably compact source, less than 7 arcminutes across—more like a star than a cloud.

These experiments showed that the sky might be full of some very interesting X-ray sources, which should appear if the sky could be scanned with a narrow beam, or even with some sort of camera which could produce a picture rather than an intensity averaged over a large field of view. What was needed first was an X-ray photometer with a narrower field of view. This would be far short of a camera, but if there were enough time to scan the whole sky with this pencil beam, a map of the X-ray sky could be drawn. The time was the problem; rockets provide only a few minutes of operation above the atmosphere. An orbiting satellite was the only solution, allowing months or years to compile a catalogue of these new sources.

Honeycombs, Pinholes, and Shadows

The first orbiting X-ray observatory was produced by NASA. It was named Uhuru—a Swahili word meaning 'freedom'—celebrating its launch in 1970 from San Marco, near Mombasa in Kenya. This is an Italian launch site, used for sounding rockets in atmospheric research. The site was chosen because it is near the equator, which is appropriate for placing a satellite in an equatorial orbit. By 1970 the new organization NASA was in full swing, launching the Explorer series of small satellites. Uhuru, originally named SAS-1 (the first Small Astronomical Satellite), carried two sets of detectors counting X-ray photons, with fields of view restricted to only a few degrees, giving a better chance of picking out individual X-ray sources. In front of each detector was a deep honeycomb grid which allowed only X-rays arriving from nearly straight ahead to reach the detector. The whole sky was scanned with these two sets of detectors—one restricted to 5° and the other to 0.5°. With the spacecraft rotating about a spin axis which moved slowly across the sky, Uhuru covered the whole sky several times during its three-year lifetime. This was one of the most rewarding space observatories, producing a catalogue of no fewer than 339 X-ray sources, shown on the map in Figure 31. Among these were several now famous individual

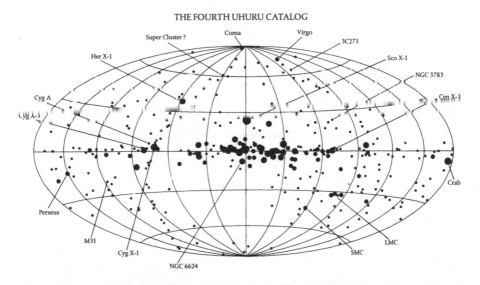

THE FOURTH UHURU CATALOG

Fig. 31. The map of X-ray sources in the Uhuru catalogue. Most sources are in the Milky Way, along the centre of this map, but many of those which are labelled are extragalactic.

objects, well-known to astronomers as Centaurus X-3, Vela X-1, Hercules X-1, and Cygnus X-1.

What was needed now was a camera capable of producing a detailed image of a patch of sky covering multiple sources, rather than a single blurred measurement of an average over an area of several square degrees. The difficulty was to build a camera without using a lens or a mirror. The first X-ray camera to be flown used a very old idea: the pinhole camera. The principle is easily demonstrated: sunlight shining through a hole in a shuttered window into a darkened room can produce a round image of the disc of the Sun on an opposite wall.

Pinhole cameras were known long before the invention of photography. In Cairo, about the year 1000, Ibn al-Haytham (Alhazen) wrote a book on optics in which he showed that the action of a pinhole camera (often now called the camera obscura) depended on light travelling in straight lines. X-rays also travel in straight lines—which is indeed the problem in making cameras for X-rays, which tend to pass undeviated through the lenses and mirrors familiar in optics. Pinhole cameras do, however, work for X-rays, using a pinhole in a sheet of a dense metal such as lead. An image can then be formed on a photographic plate, or in modern terms on the sensitive surface of a photoelectric detector operating like a television camera. Without a lens the image is faint, and a long exposure is needed. Rockets were only above the atmosphere for a few minutes; what was needed was a satellite, which could work steadily for months or years rather than minutes. The first test of an X-ray pinhole camera took place in 1974, in a small satellite known as Ariel V—one of a series happily named after Shakespeare's ethereal spirit in The Tempest.

The Ariel satellites were a joint UK/USA project which resulted from an offer by the USA for international collaboration. Ariel I was in fact the first multinational spacecraft, with the payload designed in the UK and the launch provided by NASA. Sir Harrie Massey (1908–1983), at University College London, was the prime mover, building on an existing programme of rocket research. UCL was joined by Imperial College London and the Universities of Leicester and Birmingham. This was the beginning of a very successful tradition of space research in the UK, leading to the Mullard Space Science Laboratory, established in 1966 as an offshoot of UCL. Other European countries were also aspiring to develop a programme of space research, and it became obvious that the expense and expertise could and should be shared by a consortium. Again the initiative was largely due to Harrie Massey. The European Space Research Organization (ESRO) was formed in 1964, and successfully provided satellite payloads which were launched by NASA. Europe also started to develop its own launch facilities, through an independent European Launch Development Organization (ELDO). ESRO and ELDO merged in 1975 to form the European Space Agency (ESA), with ten countries as members.

Ariel V carried two pinhole cameras. The so-called pinholes had to be large enough to admit a sufficient flux of X-rays; they were 1 cm in diameter, producing only a blurred image. The detector system itself was a new development, since it had to distinguish position as well as intensity of X-ray photons. The cameras produced images with a resolution of 10°, which is crude by modern standards but is sufficient to distinguish one source from another. With a lifetime of years rather than minutes, satellites like Ariel V have an enormous advantage over rockets in being able to look repeatedly at the same source, checking for variations or pulsations. Ariel V was able to monitor the strength of many variable sources, and discovered several pulsars, mainly those with long periods of several minutes.[1] One of these, Cygnus X-1, turned out to be a binary system, in which the source of the X-rays was a neutron star in orbit around a black hole. Ariel V also carried a spectrometer, which discovered X-ray spectral line emission from several active galaxies.

The problem with a pinhole camera is its lack of sensitivity. A single pinhole might be only a few millimetres across; but astronomers need much more sensitive telescopes, with apertures measured in metres. How could a larger aperture be used, when no focusing system existed? The solution is to replace the pinhole with a patterned mask which allows through half of the light or X-rays. The pattern can be complicated, looking like a two-dimensional barcode used to identify items in a shop (Figure 32).

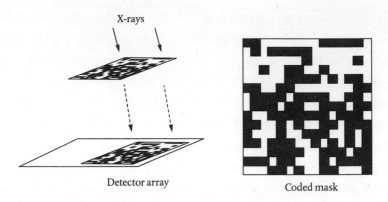

Fig. 32. A coded aperture mask used in X-ray and gamma-ray telescopes. A point source casts a shadow of the mask on a detector array. The shadows from multiple sources require decoding by computer.

A single source, like the Sun, produces a shadow of the mask, with the same pattern, which can be recognized. What happens if there are several different sources, each casting the same shadow pattern but at different locations on the detector? Surprisingly, it is still possible to decode the overlapping shadows and produce a picture of the sky. Several versions of this system, known as the coded aperture mask, were flown before reflecting telescopes for X-rays were developed. Some are still in use, as the imaging reflectors (described later) can be made to work only at the low-energy range of the X-ray spectrum, and astrophysicists insist on the importance of observing over the high-energy range (up to 100 keV) as well as in the easier low-energy range (around 1–10 keV).

The invention of a camera using such a coded mask is attributed to Hal Anger (1920–2005), and a type of mask is named after him. He was working in California in the field of nuclear medicine, where there was a need to make an image of low-energy gamma rays (around 100 keV) emanating from activated sources in the human body. Gamma rays of that energy, at the top of the X-ray range, cannot be focused, and an image can be formed only by some geometrical scheme such as a coded mask. The image, which consists of overlapping shadows of the mask, has to be recorded and analysed by computer. The detector system in Anger's camera was another novelty: an array of photoelectric detectors, each detecting the gamma rays as flashes of light generated by the gamma-ray photons as they hit a crystal of sodium iodide. The application of a coded mask to astronomy, with the theory of the analysis of the complex picture, was suggested in 1968 by John Ables, in Australia.

Unfortunately, all the coded-mask schemes are fundamentally less sensitive than a conventional telescope with a full, unrestricted aperture. The X-ray telescopes which followed them achieved complete imaging by using a novel mirror system, as described in the next section. However, this can be achieved only for low-energy X-rays, and the mask, or some other geometrical system, is still essential for high-energy X-rays and gamma rays.

The transition to the modern X-ray camera telescopes is well illustrated by the story of the European X-Ray Observatory Satellite (EXOSAT)—a major satellite planned by ESA in the early pre-camera era, but redesigned and flown as one of the new generation equipped with a focusing camera. EXOSAT was originally proposed by ESRO in 1969 as the Highly Eccentric Lunar Occultation Satellite (HELOS), intended to provide accurate locations for the many X-ray sources that were turning up in the surveys but which remained unidentified. The idea was to observe and accurately time the changes in X-ray flux from the sky as the Moon moved across it. This had already been achieved for identifying the Crab Nebula as an X-ray source, and the intention was to use the Moon to scan the sky and locate a large number of other X-ray sources. A highly eccentric orbit was to take the spacecraft a large distance away from the Moon's orbit, and track the Moon for months or years as it covered a large proportion of the sky. It always takes many years to develop a new satellite. HELOS was no exception, and, while it was developing, both the requirements and the techniques of X-ray astronomy were evolving rapidly. ESA took the opportunity of installing the first focusing cameras, with collecting areas of 10 square cm. The satellite even changed its name in the process, from HELOS to EXOSAT.

Only the choice of orbit survived the plan as it evolved; EXOSAT did not eventually observe a single lunar occultation. The orbit reached an apogee of 200,000 km, which produced a period of 90 hours—long enough for a single continued observation to reveal some remarkable periodic and non-periodic fluctuations in the intensity of several X-ray sources. EXOSAT incorporated several instruments including two imaging cameras and, for the first time, an on-board programmable digital computer. It was launched in 1983, and immediately demonstrated the power of a true imaging camera in sky surveys and in following fluctuations in individual sources.

X-Ray Mirrors

The X-ray cameras which provide the large orbiting observatories with capabilities which rival those of optical observatories were devised by Hans Wolter (1911–1978), a German physicist. Wolter intended to construct an X-ray microscope for medical research. Conventional lenses are useless for X-rays, as there is no material that refracts X-rays like visible light. Reflection is equally impossible, except for incidence on a polished metal surface at a glancing angle. The inside surface of a slightly tapered cylinder will focus X-rays arriving from a direction along the axis of the cylinder, and if the inside surface is part of a paraboloid, the focusing will be precise for a source exactly on axis. In Wolter's system the mirror, as shown in cross-section in Figure 33, has two parts. X-rays are reflected first on the inside of a paraboloid, and second on the inside of a hyperboloid. This system produces a good focus for sources located both on and for some distance off the axis, so that as in any good camera the images are near perfect over a useful field of view. This is the system known as Wolter Type 1. (Wolter Type 2 and Wolter Type 3 used different combinations of reflecting surfaces, which are unlikely to be useful in astronomy.)

From Figure 33 it is obvious that a single mirror is inefficient, missing most of the X-rays entering the telescope. However, a series of concentric mirrors, as shown, can make use of a considerable proportion of the aperture. The first large X-ray camera of this type, flown on the Einstein satellite, had a nested set of four mirrors, and the telescope aperture was

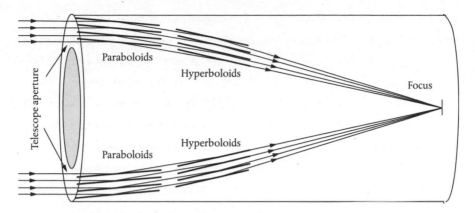

Fig. 33. The Wolter Type 1 X-ray telescope, with four concentric mirrors, as used in the Einstein and Chandra satellite observatories. The two successive sets are sections of paraboloids and hyperboloids.

60 cm. Although only a fraction of the aperture contributed to the image, the sensitivity was an improvement of more than a hundred-fold over previous telescopes. Einstein was the second, and the most successful, of a series of three High Energy Astronomical Observatories launched by NASA. A mechanism at the focal plane allowed four different instruments to be deployed, including an X-ray spectrometer. The best angular resolution, in the centre of the field of view, was a few arcseconds—admittedly only as good as a small optical telescope on a rather poor site, but good enough to discover several thousand individual X-ray sources. Several of these sources were extragalactic, including some in the Andromeda Nebula and the Magellanic Clouds.

The Wolter camera on Einstein became the pattern for several other orbiting X-ray telescopes, notably the NASA orbiting observatory Chandra. The four nested mirrors were constructed from cast cylinders of glass-ceramic—the same material as used in large optical telescopes. The inside surfaces were ground and polished, and coated with gold. For X-ray mirrors it is essential to use a heavy metal which can take a smooth polish, and several metal coatings, such as iridium and rhodium, have been used. At the very short wavelength of X-rays it is not possible to make the whole reflecting surface accurate to within one wavelength, but it must be smooth and follow the correct geometry, as any large-scale departure from a perfect shape will cause a blurring of the image.

Chandra and XMM-Newton

The X-ray observatory Chandra was launched in 1999 as one of the four Great Observatories designed by NASA to cover the electromagnetic spectrum; the others were the Hubble Space Telescope, the Spitzer Space Telescope, and the Compton Gamma Ray Observatory. Chandra was named after Subramanyan Chandrasekhar (1910–1995), the great Indian–American astrophysicist who is known particularly for his theoretical work on white dwarfs and black holes. The spacecraft was built as the Advanced X-Ray Astronomical Facility (AXAF), and was named Chandra only after it was launched successfully. It is still operating at the time of writing (2015).

The Chandra telescope (Figure 34) is 123 cm in diameter. Like Einstein, it is a Wolter telescope using four nested mirrors; in Chandra, each mirror is 2 cm thick and is accurate enough to produce very sharp images only 0.5 arcseconds across. With this configuration the effective aperture is 40 square cm,

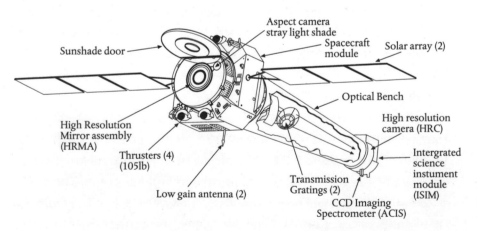

Fig. 34. The Chandra X-ray telescope. The 1.23-metre aperture is seen under the sunshade door. The long focal length of the telescope requires a spacecraft length of more than 12 metres.

which is only 3 percent of the whole. Nevertheless, the sharpness of the images ensured that the sensitivity was a spectacular improvement over Einstein, and Chandra has proved to be a continuing front-line instrument, especially for detecting and mapping faint X-ray sources. The positions of individual sources can be measured to better than 1 arcsecond. By mid-2013, Chandra had detected and located no fewer than 500,000 discrete X-ray sources. Long exposures can be used in deep surveys of selected regions, and it is claimed that the faintest sources are detected with only one photon arriving every four days.

XMM-Newton, launched by ESA in the same year as Chandra, was designed mainly for spectroscopy of faint X-ray sources, for which a large collecting area was essential. A new technique was adopted to increase the effective aperture. Instead of a single telescope with four nested Wolter reflectors, XMM-Newton has three telescopes, each 1.07 metres in diameter and with no fewer than fifty-eight nested reflectors, producing a total collecting area of 4,300 square cm. Packing this large number of reflectors into a single telescope means that their surfaces are only 2 mm apart; each is a cylindrical sheet only 1 mm thick, leaving gaps of 1 mm. Fabrication of such a structure, preserving a tolerance of around 5 microns overall, is evidently a challenge. The technique is to construct a mandrel of highly polished aluminium, precisely shaped for each layer, and to deposit the thin reflector by electroforming. This is similar to electroplating, using the mandrel as an electrode in a liquid bath. First, a thin layer of gold is deposited, which will become the reflecting surface, followed by a thicker layer of nickel to provide mechanical strength and rigidity. Parting the shell from the aluminium mandrel is achieved by cooling the whole assembly; the aluminium shrinks more than the shell, which separates with the surface, preserving the high polish of the mandrel. The assembly is held in place in the telescope by a spider-web made of a nickel alloy.

The operation of these multimirror systems is truly remarkable. X-ray photons entering a gap only 1 mm wide must be reflected twice, and only twice, in a length of around 50 cm. This means that they must arrive from a direction on axis within a very small angle, which defines the angular precision of the telescope. Within this small angle, all X-ray photons from a point-like source arrive at a point-like image at a focal plane, where an image is formed just as in an optical camera. The image is formed on an array of detectors; again as in optical cameras this is a CCD array, but adapted to be sensitive only to the higher energy of X-ray photons.

The multimirror technique is pushed even further in the orbiting X-ray telescope Nuclear Spectroscopic Telescope Array (NuSTAR, Plate 12), in which there are two telescopes each with 123 concentric shells. These are made from flat glass sheets less than 0.25 mm thick, heated and slumped on to a polished quartz mandrel. Each cylindrical shell is made from six or twelve of these glass segments; in total, for both telescopes, there are no fewer than 4,680 segments. The reflecting surface is again an innovation, using a technique that extends the useful energy range up to 79 keV—about ten times the energy available to the earlier Wolter telescopes. This is achieved by using a multilayer coating with alternate thin layers of platinum and silicon. The geometry of very thin gaps between shells demands a long focal length, which for NuSTAR is 10 metres. This is outside the scope of an ordinary spacecraft, so the detector arrays at the focus of each telescope are mounted on a mast which extends out beyond the front of the telescope (Plate 12).

The success of NuSTAR led to the design of Astro-H, which incorporates many of the new techniques and includes a long-focus hard X-ray camera which requires an extension of the spacecraft to an extra 14 metres when in space. It will be a comprehensive observatory, including soft X-rays and a gamma-ray telescope. Astro-H is being developed and launched by the Japanese Aerospace Agency (JAXA), in collaboration with NASA, the Netherlands Institute for Space Research, and the Canadian Space Agency.

The X-Ray Sky

X-ray astronomy has been opened up as a new field of astronomy by the development of these Wolter nested-mirror telescopes. A brief look at the

X-ray sky, as seen so far, shows how this new realm has touched almost every branch of astrophysics. As seen with X-ray eyes, the sky is not totally unfamiliar; the Sun and the Milky Way are still there, but the familiar bright stars of the constellations have faded into significance. The X-ray stars of the Milky Way are mostly quite different objects, energized by dynamic processes rather than nuclear energy. In other galaxies, our new telescopes are picking up X-rays from objects energized by black holes. Why is the X-ray view so different from optical astronomy? This is a question of wavelength, temperature, and photons.

X-ray wavelengths are more than a thousand times shorter than those of light, and the concept of wavelength has little significance. Instead, we think of X-rays and gamma rays arriving as photons, emitted and received as packets of energy. Only very energetic objects can emit significant amounts of X-ray photons. Such objects must either be very hot, at a million degrees or more, or they must contain very energetic charged particles, usually electrons.

The most familiar object hot enough to radiate X-rays is the solar corona, the outermost part of the Sun's atmosphere which, surprisingly, is very much hotter than the ordinarily visible surface. The strongest X-ray sources in the galaxy are, however, binary stars. The companion stars in a typical X-ray binary are very different; one is a neutron star, with a mass a little larger than the Sun, but condensed into a rapidly rotating sphere with a diameter of only about 20 km, and the other is a usually a more normal star with a huge atmosphere. These two are in close orbit round each other, so close that the enormous gravitational pull of the neutron star is sucking in the atmosphere of its companion. A stream of gas falls onto the surface of the neutron star, picking up gravitational energy. Both the gas stream and the surface of the star become hot enough to emit X-rays. The radiation from this infalling cloud and from the star surface both vary dramatically. The star surface is heated very unevenly, and as it rotates it appears to pulsate. This is an X-ray pulsar. One such was discovered in the galaxy M82.[2]

Gas clouds hot enough to emit X-rays are found in our galaxy, but there are many galaxies which are far more spectacular X-ray emitters. These are the starburst galaxies, in which there is unusually rapid star formation in clouds like the Eta Carina Nebula (Plate 11), and galaxies in collision. Plate 13 shows the two Antennae galaxies—a spectacular collision of two galaxies at a distance of 45 million light-years.

On the larger scale of clusters of galaxies, the intergalactic gas is heated as it falls in towards the centre of the cluster, reaching temperatures of 10–100 million degrees and becoming visible in X-rays. Although the mass of this gas may be many times the mass of the stars themselves in the whole cluster, it is transparent and emits no visible light and can only be seen in X-rays.

The spectrum of radiation from most of these cosmic X-ray sources is broad, like white light. At low X-ray energies there are some spectral lines, like the visible spectral lines which originate in energy levels in atoms and molecules. At higher energies the only narrow features arise in a different kind of resonance, seen in X-rays from some neutron stars. These stars have very high magnetic fields, many orders of magnitude higher than in normal stars, and the resonance arises from electrons gyrating round such a high magnetic field. The high frequency of their radiation appears as an X-ray spectral line. NuSTAR has detected such a spectral line with an energy of 31 keV, from a binary X-ray source in the Large Magellanic Cloud.[3]

Finally, the X-ray sky contains many objects which are variable on a variety of time-scales. Many of the younger pulsars, which are usually observed by radio astronomers, also radiate gamma rays and X-rays. There are also transient sources—short bursts of X-rays and gamma rays—whose origin is still unknown but may be supernovae in very distant galaxies. Hunting for the transient sources is particularly demanding, requiring a telescope which is sensitive over a large area of sky.

The X-ray telescopes described in this chapter are variously adapted for observing this wide variety of objects. Chandra has the most precise imaging system, providing 0.5-arcsecond accuracy in a very large catalogue of individual X-ray sources. XMM-Newton has a larger collecting area, and is particularly useful for spectroscopy. Both these operate at the low-energy end of the X-ray spectrum, and NuSTAR extends the energy range up to

79 keV, though with less precise imaging. A different approach is needed for the transient sources, and here the key spacecraft is Swift—an orbiting multi-wavelength observatory with gamma-ray, X-ray, and ultraviolet telescopes.

Swift

Refreshingly, the name Swift is not an acronym. This spacecraft, operational since 2005, is designed to search for gamma-ray bursts—a phenomenon which has defied explanation for many years. Apparently they occur randomly at any direction in the sky, which suggests that they originate not in our galaxy but at very great distances, in which case they must be very energetic indeed, caused by the most powerful explosions in the entire Universe. Swift is required to locate the origin of a burst which may be only a few milliseconds long, and from any direction in the sky.

The bursts are detected by a telescope covering a wide angle of sky, and working at the high X-ray or low gamma-ray energy range of 15–150 keV. This is the Burst Alert Telescope (BAT), which measures the burst position to a few arcminutes. When a burst is detected by the BAT, the whole spacecraft can be slewed into that direction within a few seconds to allow its X-ray and optical telescopes to look for any afterglow. Furthermore, the position of the burst is transmitted to a ground station and distributed to observatories all over the world. The name Swift aptly refers to the speed and agility of this operation, like the bird of the same name. The X-ray telescope is another Wolter Type 1, with twelve nested segments. The BAT measures the burst position to a few arcminutes, and as soon as the burst source is in the field of view of the X-ray telescope it can provide a direction accurate to a few arcseconds.

The BAT operates at high photon energies of 15–150 keV, beyond any possibility of focusing optics. Instead, it uses the coded aperture mask technique introduced in early X-ray observations, before the development of Wolter grazing incidence telescopes. The mask has to be large, with many tiles, to cover such a large area of sky; the mask on Swift has 54,000 lead

Fig. 35. The coded mask at the aperture of the Burst Alert Telescope on the Swift spacecraft. The mask has 54,000 lead tiles, each 5 mm square. Randomly arranged, they cover half of the telescope aperture.

tiles randomly arranged on a mask covering 2.7 square metres (Figure 35). The detector array has 32,768 CdZnTe detectors assembled in 256 modules of 128, all individually addressed. The on-board computer is required to recognize a shadow pattern of the coded mask anywhere on this array, check that the source is transient, and find a location. Using the BAT, Swift detected and located more than five hundred gamma-ray bursts in its first five years of operation.

X-Ray Prospects: ATHENA

X-ray astronomy is well provided for with the present array of Newton-XMM, Chandra, NuSTAR, Swift, and Astro-H. Every spacecraft, however, has a limited life, and the telescope designers are already at work on the

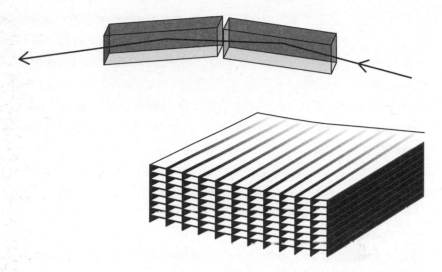

Fig. 36. Silicon pore optics, as proposed for the ESA X-ray spacecraft ATHENA. Grooved silicon wafers 8 cm square are stacked to form box-like elements, of which 250 are needed for each of the two telescopes.

next generation. ESA is developing a proposal for an X-ray telescope, using new techniques for X-ray optics and detectors and providing a sensitivity a hundred times that of the existing telescopes. The Advanced Telescope for High Energy Astrophysics (ATHENA) is scheduled for launch in 2028. Working in the energy range 0.3–12 keV, it will have two telescopes—one for spectroscopy and one for wide-field imaging. Each has an effective area of around 1 square metre. The aim is an angular resolution of 5 arcseconds. This is achieved using a new version of the Wolter Type 1 grazing incidence system, known as silicon pore optics. Figure 36 shows the basic element: a nested array of channels, with two layers of channels forming the two elements of a Wolter telescope. The novel feature is that each element, only 1 mm across, is itself a Wolter telescope; the array comprises 1 million of these channels. The array is fabricated from layers of thin silicon plates with the channels cut as grooves, and the reflecting surfaces are coated with gold.

Spectroscopy demands a detector in which each pixel element can measure energy rather than simply signal the arrival of a photon. In ATHENA this will be achieved by using the same bolometer technique as in the infrared telescopes described in Chapter 4. This is the Transition Edge Sensor, a device that must be kept at a temperature below 1K; fortunately, this very low temperature is now achievable with on-board refrigerators rather than with an expendable tank of liquid helium. The complexity of an array of such detector elements restricts their number to a comparatively small array of 1,000 elements. The detector array for the wide-field imager uses transistors; half a million of these can be packed closely together and read out sequentially as in a CCD. In the X-ray region around 1 keV they have a very high quantum efficiency, detecting 98 percent of the incident photons.

The complexity of such a spacecraft, with its two telescopes and sophisticated detector arrays, demands a long lead time, during which every aspect of spacecraft and instrument engineering is made as near perfect as possible. Projecting the future of X-ray astronomy beyond the launch of ATHENA can only be a speculation. Astronomers are never satisfied, and they will doubtless be pressing for newer capabilities, including the challenge, so far almost unmet, of measuring the polarization of X-rays.

7

GAMMA RAYS AND
COSMIC RAYS

We now move to the highest energies in the electromagnetic spectrum, above X-rays, and extending over a very wide range to energies as yet unattainable in any terrestrial high-energy physics laboratory. This is the gamma-ray region of the spectrum, with the same high energies as cosmic rays, the particles which continually bombard Earth and are equally interesting to astronomers. Low-energy gamma rays were first observed and named as one of three types of radiation from naturally radioactive materials such as radium. These three types were categorized according to their power of penetrating through materials such as thin sheets of metal. Alpha particles were the least penetrating; these turned out to be atomic nuclei, and are now identified as helium nuclei. Beta particles are electrons, which penetrate further. The most penetrating are the gamma rays. Although these were originally thought of as material particles, gamma rays are in fact very energetic photons, discrete packets of wave energy behaving like the photons which have become familiar in visible light. Gamma rays are as much a part of the electromagnetic spectrum as radio waves, for which we only rarely think in terms of photons.

Waves and Photons

What is hard to grasp is the very wide range of wavelengths and energies across the electromagnetic spectrum. Radio waves have an easily visualized wavelength of metres down to millimetres. The wavelengths of visible light are more than 1,000 times shorter, X-rays another 1,000 times shorter, and even the longest-wavelength gamma rays yet another hundred times shorter still. The corresponding photon energies follow the same ratios, with light conveniently in the centre of the spectrum with energies close to 1 eV (the typical energy of an electron in an LED, the ubiquitous light-emitting diode). For the longer-wavelength radio the photon energies are thousands or millions of times smaller. The lowest-energy gamma-ray photons have energies of 100,000 eV (100 keV), and the energies of gamma rays which have been observed to arrive on Earth extend very much further again, by another factor of at least 100 million (Figure 37).

Throughout the more familiar parts of the spectrum, from radio through light and X-rays, we have found that there is a wealth of information about our Universe. We have found how to devise and build telescopes to collect and interpret the streams of radiation at these longer wavelengths. Is there also a stream of gamma rays arriving from the cosmos, and can we find ways of capturing and interpreting these energetic photons?

The gamma-ray sky is very different from the familiar night sky. Even at X-ray wavelengths there are some familiar objects, such as the solar corona, which are hot enough to emit detectable radiation. Particles which are sufficiently energetic to radiate gamma rays are found only in extreme situations, such as in a supernova explosion. Gamma-ray telescopes also look very different from optical telescopes, since there is no possibility of focusing with lenses or mirrors to create an image. If we want to know

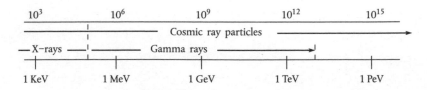

Fig. 37. High-energy radiation from the cosmos. Gamma rays extend from X-ray energies to at least 10^{13} eV (10 TeV), while cosmic-ray particle energies reach to over 10^{20} eV.

where gamma rays come from, we must detect the individual photons, measure their energy, and track their direction of arrival.

Gamma-ray astronomy has to contend with a confusing background of cosmic rays. These are real material particles, mostly protons, showering down on the Earth in a continuous rain at a rate some thousands of times greater than the gamma-ray photons; furthermore, they cover a similar range of high energies, and may overwhelm any detector aiming to detect only the photons. The cosmic rays themselves are of great interest, their origin being even less understood than that of the gamma rays. Neither the gamma-ray photons nor the cosmic-ray particles can penetrate the Earth's atmosphere, and any direct detection has to be achieved by a spacecraft. Distinguishing the two in the spacecraft requires techniques that resemble those of nuclear physics rather than a conventional optical observatory. Less directly, both cosmic rays and gamma rays can be detected from the spectacular showers of charged particles, mainly electrons and protons, which result when they encounter the atmosphere. Fortunately, the showers produced by gamma-ray photons are appreciably different from those produced by the cosmic rays. An optical telescope system can be used to photograph the light emitted by the shower particles, and the gamma-ray events can be distinguished by the characteristic shape of their showers.

Gamma-Ray Bursts

The first indication that something entirely new about the cosmos might be revealed by gamma rays came in the 1960s as an unexpected result of the Cold War, when the USA and the USSR were competing to develop nuclear weapons. When the Partial Test Ban of 1963 marked the end of nuclear tests in the atmosphere, the USA launched a series of satellites which would monitor compliance with the ban. These would watch for a characteristic burst of gamma rays which would reveal any nuclear test in the atmosphere. The existence of these Vela satellites was secret, so when, soon after the first pair was launched in 1963, they started to detect frequent bursts of gamma rays with characteristics unlike those of nuclear explosions, nothing was revealed to the scientific community. The first publication came in 1973, and it was not until much later again that the origin of the bursts was found. The Vela detectors were unable to locate the source accurately, and the longest bursts lasted only a few minutes, giving little time to alert astronomers to search for anything unusual. More than two thousand bursts had been reported before there was any clear indication of their origin. The breakthrough came in 1997, when an Italian–Dutch X-ray satellite, BeppoSAX, detected a continuing afterglow lasting for some hours after a strong gamma-ray burst, and was able to pinpoint its position with an accuracy of 1 arcminute. Rapid world-wide communication alerted optical observers, and with great good fortune a visible afterglow in a faint very distant galaxy was seen by the William Herschel Telescope on La Palma.[1] An afterglow from another burst in the same year was seen by several telescopes, including the Keck 10-metre in Hawaii, which found that the spectrum of the fading light[2] showed spectral lines at the astonishingly large redshift of z = 0.835, indicating a distance of 6 billion light-years. For a source at that distance the energy in one single short burst was equal to the total energy output of a star like the Sun for the whole of its lifetime.

Gamma-ray bursts originating in such distant galaxies are now known to be a normal occurrence. They were routinely observed by the first large spacecraft devoted to gamma-ray astronomy. This was the huge Compton Gamma Ray Observatory (CGRO), a 17-tonne spacecraft which was launched from the Space Shuttle in 1991. The CGRO detected bursts at the rate of around one per day throughout its lifetime of nine years. These appeared to arrive from anywhere in the sky, without any concentration towards the centre of the Milky Way, confirming the idea that they originated in distant galaxies. A very large burst observed in 1999, which was followed up by ground-based telescopes, proved to be from a galaxy with red shift z = 1.6, indicating an even more distant origin.

The burst detector systems on the CGRO extended the energy spectrum from bursts up to 2 MeV (2×10^6 eV). The detectors in this observatory, and in most spacecraft of this era, used a large crystal of sodium iodide, which responds to a gamma ray by producing a flash of light. Eight of these scintillation detectors were distributed at the corners of the spacecraft, and a gamma ray which passed through two of them provided a rough measurement of the direction of its origin. A more accurate measurement needs to follow the track made by a photon as it passes through a much larger volume. This was first achieved in Cos-B, a European spacecraft launched by ESRO in 1975. In Cos-B, the track of a gamma ray could be traced as it passed through a series of metal grids in a large tank of gas, leaving a trail of ionization. It also triggered a pulse of high voltage between the grids, which made a spark along the ionized gas of the gamma-ray track. This could then be seen by an array of photomultiplier tubes, producing a three-dimensional view of the track. With this 'spark chamber', Cos-B achieved a detector area of 50 square cm and an angular accuracy of around 2°.

A larger spark chamber—the Energetic Gamma Ray Experiment Telescope (EGRET)—was installed on the CGRO. As always in telescopes for any wavelength or energy range, the collecting area is very important; and for gamma rays it is vitally important, since higher-energy gamma rays are rare. Early gamma-ray telescopes detected gamma rays with energies around 1 MeV; but what was needed was a detector that would reach energies in the GeV region, some thousands of times more energetic but at a correspondingly lower rate. The spark chambers in the CGRO and Cos-B showed how this could be done. They had already detected a substantial catalogue of gamma-ray sources, and produced a gamma-ray map of the Milky Way. The central idea of the spark chamber was eventually developed into the Fermi Large Area Telescope (LAT), launched in 2008, which achieved a collecting area of 1 square metre and a directional accuracy of 1 arcminute. It also timed the arrival of each individual photon to an accuracy of 1 microsecond.

The cosmic-ray bursts themselves, which started this line of development, are now monitored by a dedicated spacecraft Swift, which has been operating since 2005 (Chapter 6). The three instruments on Swift present a nice combination of telescope techniques: the burst detector uses a large coded mask, developed from the ideas in some early X-ray telescopes, while lower-energy X-rays are imaged by a Wolter Type 1 camera with twelve nested mirrors. The optical telescope uses a design adopted as an optical monitor for Newton-XMM. This remarkable assembly of new techniques, with the capability of such a rapid response, has opened up the whole new field of transient high-energy bursts. Up to 2014, Swift had detected more than a thousand gamma-ray bursts, of which 90 percent were found to have X-ray components. The optical spectra, when seen, showed spectral lines with very large redshifts, indicating an origin in very distant galaxies in the young Universe. In 2008, Swift initiated a spectacular sequence of observations[3] of a burst, providing a location accurate to 2 arcseconds, which was transmitted to the Hubble and Chandra space telescopes, the VLA radio telescope, UKIRT for infrared, and seven of the largest optical telescopes. Within a few hours, UKIRT found that the spectral feature known as the Lyman-α edge was visible at a wavelength redshifted by no less than eight times, and the VLT (see Chapter 3) measured this redshift at $z = 8.26$. This placed the source at a greater distance than any known galaxy or quasar. The burst is believed to indicate the collapse of a huge star, one of the earliest formed soon after the era when the Universe first became visible (see Chapter 11). At the time of the burst the apparent age of the Universe was 630 million years—only 4.6 percent of its present age.

AGILE and the Fermi LAT

Building on the experience of the CGRO and Cos-B, and incorporating the expertise of high-energy physics laboratories, two new gamma-ray telescopes were launched: Astro-Rivelatore Gamma a Immagini Leggero (AGILE) in 2007, and the Fermi Large Area Telescope (LAT) in 2008. Both of these included a burst monitor, which in the Fermi LAT is detecting around two hundred bursts per year at energies of around 1 MeV. The large number of burst detections, and an adequate measure of the direction of arrival, showed clearly that their sources are distributed uniformly over the sky, and it is now accepted that these are all from very distant galaxies. The most important innovation in AGILE and the Fermi LAT was a development of the spark chamber using a new type of detector. This extended the energy range to 300 GeV, and enabled the location of some hundreds of individual sources with an angular accuracy of a fraction of a degree. Fermi LAT also measured the time of arrival of each gamma-ray photon, and transmitted it to the tracking stations with microsecond accuracy. This proved to be the most important and productive capability of this space observatory, opening a new field of research into gamma-ray pulsars.

Pulsars were discovered by their radio emissions, at the opposite end of the electromagnetic spectrum from gamma rays. These pulsating sources of radio were identified as rapidly rotating neutron stars emitting a narrow lighthouse beam of radio waves. As early as 1975, two of them (the Crab and Vela pulsars) were found to emit pulsed gamma rays, and the possibility that there might be many others stimulated the inclusion of an accurate timing system on Fermi LAT. The key in detecting any radiation from a pulsar is an accurate knowledge of its pulse periodicity, allowing a long sequence of weak pulses to be added. For gamma rays, the situation is particularly difficult, as the very energetic photons arrive only very rarely, at a rate much slower than the pulse periodicity. Despite this drastic undersampling, Fermi LAT has detected gamma rays from many of the known pulsars, and has gone on to achieve the seemingly impossible task of discovering pulsars which only emit pulses of gamma rays, and have no radio counterparts to give the pulse periodicity. To see what is involved in such an achievement, we should look in more detail at the design and operation of the spacecraft and its gamma-ray telescope (Figure 38).

The heart of the telescope is a vacuum chamber of about 1 metre cube. In this chamber a gamma-ray photon encounters a series of thin metal sheets, made of a heavy element such as tungsten. The photon may pass through several of these, but at one of them the photon will hit a heavy atomic nucleus and convert into a pair of particles: an electron and a positron. These diverge sufficiently to produce a V-shaped pair of tracks as they pass through a series of crossed layers of silicon-strip detectors which act like

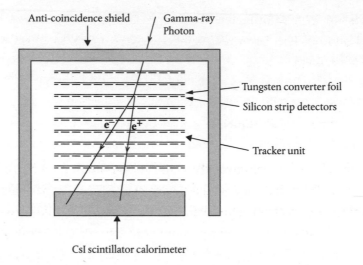

Anti-coincidence shield

Gamma-ray
Photon

Tungsten converter foil

Silicon strip detectors

Tracker unit

CsI scintillator calorimeter

Fig. 38. The detector system of the Fermi Large Area Telescope (LAT). A gamma-ray photon converts to an electron–positron pair at one of several sheets of tungsten foil. The pair are detected separately in the silicon-strip detectors, and the total energy is measured by their scintillation light in the caesium iodide crystal.

the CCDs familiar in optical cameras. The mean line between the two tracks gives the direction of arrival of the gamma-ray photon. This is a direction in relation to the spacecraft; to find the direction in the sky, the attitude of the spacecraft itself at the time of arrival must be known. All this is achieved to an accuracy of a fraction of a degree, including allowing for any rotation of the spacecraft. The position of the spacecraft is found from the same system, GPS, that is used for the satellite navigation and mapping system familiar in mobile vehicles. GPS also provides very precise timing; the arrival time of each photon is measured to an accuracy of 1 microsecond.

A general difficulty in all gamma-ray telescopes is that the gamma-ray photons arrive amidst a continuous rain of energetic material particles, and must be distinguished from them. These are the cosmic rays, which arrive at a rate some thousands greater than the photons. Some of these cosmic rays are absorbed in the outer shell of the spacecraft, while others must be recognized by their different behaviour at the metal sheets and be rejected in the analysis. Finally, the pair of particles generated by a genuine gamma ray ends up in a calorimeter, which measures the total energy of the original photon. The Fermi LAT is indeed a very complicated and sophisticated instrument.

Picking up the Pulses

When the pulsars were discovered and named as pulsating sources of radio, it seemed unlikely that they would be detectable in any other part of the electromagnetic spectrum. However, the description of a pulsar as a rapidly rotating and highly magnetized star suggested that some form of high-energy dynamo was acting which could accelerate electrons and positrons to extremely high energies, so that they might emit gamma rays. The first target to be searched was the Crab pulsar—a very energetic radio pulsar which was also known to emit pulses of visible light. Already in 1971 a balloon-borne spark chamber had been used to show that the Crab Nebula contains a source of gamma rays with energy greater than 35 MeV.[4] This was later shown to be the pulsar. The first spacecraft to discover gamma rays from a pulsar was Small Astronomy Satellite 2 (SAS-2), which observed the Vela pulsar, another strong radio pulsar.[5]

The major gamma-ray observatories Cos-B and CGRO then showed the way with a series of detections of sources which were clearly associated with pulsars, and also some dozens of other individual gamma-ray sources which could not be related to anything observed previously.

Pulsar radiation comes in the form of a very characteristic signal—a pulsation at a precise rate, which for the Crab pulsar is 30 per second, or a rate of one pulse per 33 milliseconds. Radio observations are routinely made nearly every day to measure the precise rate of pulsation, which is observed to be slowly decreasing, and the time of each pulse can be predicted days in advance. With this background, the gamma-ray astronomers should have an easy task. The gamma-ray photons should arrive in synchronism with the radio pulses. However, gamma-ray photons are so rare that only one could be expected to arrive in a very small proportion of the expected times. Patience will be rewarded, or as the observers would say, a long integration time is needed. Using the radio prediction, the observers can look for any gamma-ray photon arriving at any of millions or more of the expected pulse times. The most likely candidates for such a search are the youngest pulsars, since these are the most energetic. The Crab pulsar is among the youngest; it was born in a supernova explosion in the year 1054. Other known pulsars to be searched in gamma rays had periods from a fraction of a second up to several seconds. The Fermi LAT scans slowly over the sky, and at any time it accepts gamma rays from a wide field of view. The recorded photons may come from many pulsars simultaneously. For each photon a position and a time must be recorded, and any coincidence with a known pulsar will emerge only during later analysis. The timing has to be precise to a microsecond, including allowances for propagation time from the spacecraft to the observer. It is a measure of the efficiency of the whole system that by mid-2013 Fermi LAT had detected no fewer than forty-two of the known radio pulsars, and had extended its search to another class, the millisecond pulsars. These have much shorter periods, down to less than 2 milliseconds, but the precision achieved in the timing allowed even forty of these to be detected.

The strangest aspect of these detections is the way in which the radiation reaches the telescope in discrete pulses rather than as a continuous flow of energy. Photons may arrive at a rate of one every few hours, at apparently random times, but only within the regular pulse intervals. The low rate means that long runs of observing are needed before a detection can be certain. It also means that the detection must be based on a precisely known period, with a precise timing based on full knowledge of the position of the spacecraft. Some detections become possible only by compiling a whole year or more of observations and relating them to the known radio periodicity of the pulsar. This remarkable achievement was surpassed when Fermi LAT started to discover pulsars which had never before been detected, and whose periodicity was therefore totally unknown.

It seemed impossible to detect the characteristic pulsar pattern without knowing a precise period, but that is exactly was achieved by the end of

2014 for no fewer than forty more pulsars which had never been observed by radio astronomers. These were in fact already known as gamma-ray sources, among a list of sources discovered as discrete objects and reasonably accurately located by Cos-B. They were suspected of being pulsars, but there seemed no chance of finding a periodicity. Using photon arrival times collected for as long as a whole year, and analysing the intervals between them, most of these gamma-ray sources proved to be pulsars. The total number of pulsars in the growing catalogue from Fermi LAT reached 140 by the end of 2014, and was still rising.

The big step forward in gamma-ray telescopes demonstrated by Fermi LAT was the use of silicon detectors in place of the spark chambers of the pioneering SAS-2 and Cos-B spacecraft. These solid-state detectors respond well to the lower range of gamma-ray energies, which might be called the GeV range (1,000 million or 10^9 eV). The energy spectrum does, however, extend far beyond this range.

We need new designations for the much higher energies, beyond the range of spacecraft, which can be detected only by their effect on the terrestrial atmosphere. Mega (10^6) and giga (10^9) are becoming part of common speech, but we need to go much further, with tera (10^{12}), peta (10^{15}) and exa (10^{18}), to characterize the ultra-high energy range of cosmic rays. Proceeding up this energy range, gamma rays become increasingly rare. Even in the gigavolt range, the continuous rain of material particles, the cosmic rays, whose origin is unknown, outnumber the gamma-ray photons by a factor of some thousands. In this ultra-high energy range, both the gamma rays and the cosmic rays can be detected only by observing the spectacular effects when they arrive in our atmosphere.

Cosmic Rays

Most school laboratories have a demonstration of a gold-leaf electroscope—the earliest and simplest electrical measuring instrument—which is used to detect static electricity. When it is charged, a small piece of gold leaf stands up like brushed hair on a dry day. A familiar problem is that an electrically charged electroscope, however well constructed and insulated, tends to lose its charge to the surrounding air. The reason is that the atmosphere is slightly electrically conducting, because a small proportion of its atoms and molecules are ionized. In 1911 Victor Hess (1883–1964), an Austrian physicist who was interested in radioactivity, set out to find the origin of this ionization. The theory was that the ionizing radiation arose from radioactive minerals; if so, the effect would be less at large distances from the surface of the Earth. In August 1912 Hess took an electroscope to an altitude of 5 km, in a dangerous flight using a hydrogen-filled balloon (Figure 39), and found that

Fig. 39. Victor Hess at a balloon launch, Vienna, 1911. Hess discovered ionization in the upper atmosphere, caused by cosmic rays.

the electroscope discharged faster at high altitude, even though the air density was lower. He had found that the ionization was coming from some energetic source outside the Earth: he had discovered cosmic rays. The centenary of his definitive balloon flight was commemorated with a monument at the landing site at Bad Saarow-Pieskow, in Germany.

Hess's achievement was eventually recognized in 1936 by the award of a Nobel Prize. By this time, cosmic rays were well recognized as a source of interesting high-energy particles. The tracks of individual particles could be seen in a cloud chamber, invented by C. T. R. Wilson (1869–1959), in which the tracks of ionization showed up as lines of condensed water droplets. The particles which could be detected in this way were part of a huge shower of elementary particles created by a single very energetic cosmic-ray particle colliding with atoms in the atmosphere. The cloud chamber was displaying a variety of particles, mainly electrons, but among these were positrons, seen for the first time by Carl Anderson (1905–1991).

Cosmic rays can be said to have generated a shower of Nobel Prizes: Anderson shared one with Hess in 1936; Robert Millikan (1868–1953), who invented the term 'cosmic rays', was awarded one in 1923; C. T. R. Wilson was awarded one in 1927 for the cloud chamber; and in 1948 Patrick Blackett (1897–1974) was awarded one for investigations using cloud chambers to study the various different types of elementary particles in cosmic-ray showers.

Air Showers

The energies of individual cosmic rays which generate air showers extend far beyond the energies which can be reached in the largest laboratory accelerators. Direct observation of the primary particles themselves can be achieved only above the atmosphere, where apparatus in spacecraft has shown that almost all are atomic nuclei, starting with protons (hydrogen nuclei) and extending through the whole periodic table up to uranium. Only about 1 percent are electrons. The most energetic particles are naturally the most interesting, but these are so rare that there is no chance of observing them in the small detectors that can be flown in a satellite. Fortunately, the atmosphere itself can act as a detector; an energetic cosmic ray reaches the upper atmosphere with explosive energy, colliding with atoms and creating a shower of particles which grows as a cascade containing millions or even billions of particles—mainly elementary particles such as pions, muons, and electrons. The shower makes itself observable in three ways. Some of the shower particles reach the surface of the Earth, and can be detected directly, as was achieved with Wilson's cloud chamber. Second, the air itself will respond to bombardment by emitting a flash of light; this is fluorescence from atmospheric nitrogen atoms. Finally, very energetic charged particles moving through the atmosphere at nearly the speed of light themselves radiate light and radio waves—a process known as Cherenkov radiation.

Cosmic-ray air showers were discovered in 1939 by Pierre Auger (1899–1993), a French physicist well known for creativity in many fields of science. He observed the arrival of bursts of particles in a pair of detectors at some distance apart, showing that the bursts were simultaneous and must be part of a huge cloud 1 km or more across. He reckoned that a big shower must contain at least a million charged particles, and that the single particle which started it had an energy of at least 10^{15} eV (1,000 GeV), millions of times larger than any other particle or photon known to be arriving from the cosmos. Furthermore, any small detector within an area of 1 km square would detect the shower and any others of higher energy. His discovery was the inspiration for the construction of several large and very unconventional telescopes dedicated to measuring the energies and arrival directions of the very highest-energy cosmic rays. The largest of these is appropriately named the Pierre Auger Observatory.

Cosmic-ray particles with the comparatively low energies of several million eV arrive at the rate of several hundred per square metre every second. The rate falls rapidly with increasing energy; above 1 EeV (exa = 10^{18} eV) the rate is one per week per square km. Cosmic rays with even higher energy certainly exist; a few have been detected with energy 10^{20} eV, arriving at the rate of one per century per square km. A very large array of particle detectors, covering a very large area, is needed to catch such rare events. When they do arrive, their energies are measured by their extent on the ground; at the highest energies, a shower can cover 16 square km. To catch these rare events, the Pierre Auger Observatory has 1,600 detectors spaced 1 km apart, covering an area of 3,000 square km—thirty times the area of Paris.

Cherenkov Radiation: the Blue Glow

Small nuclear reactors are often surrounded by a tank of water which acts as a shield, absorbing dangerous high-energy gamma rays generated in the reactor. The gamma rays create pairs of electrons and positrons in the water, which glows with a blue light through a process known as Cherenkov radiation,. This form of radiation is named after the Russian scientist Pavel Alekseyevich Cherenkov (1904–1990), who in 1937 observed it in a bottle of water close to a radioactive material. Remarkably, this effect of a rapidly moving charge had been investigated theoretically in 1888 by the brilliant and eccentric Oliver Heaviside (1850–1925), who is mainly known for his work on telegraphy. The definitive analysis was done, also in 1937, by Igor Tamm and Ilya Frank. The three Russians shared the 1958 Nobel Prize for the discovery and the theory. As we shall see, their work contributed to observations of cosmic ray showers, but it would perhaps be too much of an extrapolation to include them in our 'shower' of Nobel Prize winners more directly connected with cosmic rays.

The principle behind Cherenkov radiation is that an energetic particle can move faster than the local velocity of light, which is possible because the light velocity is lower in a refractive medium such as glass or water. The effect is like the sonic boom of an aircraft moving faster than the speed of sound. It happens easily for elementary particles such as electrons and positrons, which in a cosmic-ray shower can be moving faster than 99.99 percent of the free space speed of light, while in air light can travel more slowly; at normal air pressure at ground level the speed of light is reduced to around 99.97 percent of its value in a vacuum. This means that a cosmic-ray shower should arrive at ground level accompanied by a flash of light. As we will see later, the flash of light in the air is a potent means of detecting showers, but Cherenkov radiation is also exploited in the detection of the particles themselves as they arrive at ground level. As in the original discovery, this is achieved in tanks of water, using fast photoelectric devices that can detect the flash from the arrival of individual particles.

In the Pierre Auger Array (Figure 40), each of the 1,600 detector stations consists of a tank containing 12 tonnes of pure water, in complete darkness except for the flashes of Cherenkov light from cosmic-ray showers. These flashes are detected by a sensitive photoelectric tube, which transmits to a central station the intensity of the flash and its time of occurrence. The extent of the shower and the total count of arriving particles give the energy of the primary cosmic-ray particle, and the small differences in arrival time across the array of detectors give the direction from which it

Fig. 40. Part of the Pierre Auger Cosmic Ray Observatory in Argentina. There are 1,600 of these detector stations, spread over an area of 3,000 square km.

arrived. The spacing of the detector stations is close enough for up to a hundred to respond to a single high-energy shower.

The Fly's Eye

At four locations around the edge of the Pierre Auger Array there are telescopes designed to record the flash of fluorescence excited in nitrogen molecules in the atmosphere. Each station has six telescopes covering 30° by 30°—a very wide field for any camera, especially for one with sufficient size to collect the faint ultraviolet light and outline its development in the sky. The basic design is a Schmidt (see Chapter 2), with a 2.2-metre aperture and a large 3.6-metre spherical-surface main mirror. At the focus there is an array of 440 photomultiplier tubes, which can respond very rapidly to individual photons. The six telescopes together cover 180°, so that any flash in the sky above the array of particle detectors can be seen. A complete picture of the fluorescence from a single shower is particularly valuable in determining the energy of the original cosmic-ray particle, which is found from the total energy in the fluorescence. The height at which the shower reaches a maximum density is also a good indicator of the original energy, since the larger showers penetrate further into the atmosphere. A combination of the surface detectors and the fluorescence detectors can provide sufficient information about the way in which the shower develops that it becomes possible to deduce the mass of the original particle as well as its energy.

The design of these fluorescence detectors derives from a series of experimental systems known as the Fly's Eye, built in various versions from 1993 by the University of Utah in the USA. One of the first detector stations had no fewer than sixty-seven individual reflecting telescopes, each of 1.5-metre aperture, arranged to cover almost the whole sky. Analogously with the development of animal eyes, subsequent designs moved away from the close similarity to the multiple simple elements of the fly's eye and concentrated more angular resolution and more pixels into a smaller number of cameras, each covering a larger field of view. Although they can operate only when the sky is dark and clear of clouds, they are invaluable in analysing the highest-energy showers.

The Pierre Auger Observatory was built by a large international consortium, combining the pioneering work of James Cronin at the Fly's Eye, and Alan Watson of Leeds University, who developed the surface particle detectors and built a large array of them at Haverah Park. Their aim was to extend the known energy spectrum of cosmic rays to above 10^{19} eV, and to find the nature of the primary particles. Their origin remains almost completely obscure, apart from a strong indication that they originate in active galactic nuclei. The array can give the direction of arrival, but this may be very different from the direction of origin, since the magnetic field of the Milky Way is sufficiently large to deviate a charged particle through a large angle. Only the highest-energy particles are relatively undeviated, and for these an origin in our galaxy should eventually show up as a distribution over the sky resembling the familiar bright streak of the Milky Way.

Sorting Out the Showers

Cosmic rays are material particles; gamma rays are photons. But the high-energy members of these two equally remarkable categories are hard to distinguish, since both are detectable only by the showers they create as they hit the Earth's atmosphere. Sorting out the gamma rays from the much more numerous cosmic rays is vital if we are to understand where they come from. This seems impossible if we look only at the shower particles which reach the Earth's surface; only if we can look at the details of the early development of the shower high in the atmosphere can we make the distinction. Fortunately, the particles in the shower radiate a flash of light by the same Cherenkov process as in the water tanks of the Pierre Auger Array, and the shower can be photographed, even in the thin atmosphere at tens of kilometres above the Earth's surface. What is needed to record this brief flash is a camera with multiple photomultiplier tubes and a wide field of view. The Fly's Eye was intended to record the scintillation of air molecules, but a much larger camera is needed to collect the faint flash of Cherenkov light from the particles of the shower themselves. It need not have the pinpoint accuracy of a conventional telescope, but only enough to delineate the outline of the developing shower. Fortunately, gamma-ray and cosmic-ray showers develop in different ways, producing different shapes and at different heights in the atmosphere, so that the comparatively rare gamma rays can be distinguished.

At the top end of the energy range, the energies of the cosmic-ray particles and the gamma-ray photons are incredibly large, a single photon or particle reaching the same energy as a rifle bullet or a tennis ball served by a Wimbledon champion. Both photon and particle create secondary particles and photons, but the gamma-ray photon continues on its track further than the material particle, producing an elongated shower. The photograph of the shower has to be good enough to show the shape of the whole shower, and to determine the direction of arrival of the photon or particle (usually a proton) that initiated it. A stereoscopic view is almost essential. A site with clear skies is needed, though a high-altitude site is not essential: the telescope is required to photograph the atmosphere, not to rise above it. Several such telescopes, aimed at ultra-high-energy gamma rays rather than cosmic rays, are in operation. Plate 14 shows a typical example: the Major Atmospheric Gamma-Ray Imaging Cherenkov (MAGIC) telescope on La Palma, in the Canary Islands.

Two identical telescopes, each of 17 metres diameter, are used to produce a stereoscopic view of the shower. Since there is no requirement for great angular accuracy, the telescope mirror is constructed economically from identical segments, each 50 cm across and with a simple spherical surface. At the focus is an array of 496 photomuliplier tubes. The lightweight carbon-fibre structure of the whole telescope enables it to move rapidly to any part of the sky in response to any report of a gamma-ray burst. A similar stereoscopic system telescope comprising four 12-metre and one 28-metre reflectors is in operation in Namibia (the High Energy Stereoscopic System, HESS), and another is at Mount Hopkins, in Arizona (the Very Energetic Radiation Imaging Telescope System, VERITAS).

These air-shower telescopes detect gamma rays with a wide range of energies, from 25 GeV up to 30 TeV (the upper limit is simply the extreme rarity of events with higher energy). More than a hundred discrete sources have been detected, starting with the Crab pulsar and extending out to some very distant quasars. These are the compact centres of active galaxies, apparently radiating an energy greater than a hundred times the total output of our galaxy, and seen from a distance of several billion light-years.

8

THE NEW RADIO WINDOW

Radio from the Sky

The name Hertz is familiar as the scientific unit of frequency, replacing the cumbersome 'cycles per second'. It was Heinrich Hertz who in 1887, inspired by James Clerk Maxwell's theory of electromagnetism, first demonstrated radio waves transmitted several metres across a laboratory. The idea that radio waves might be received from the Sun was soon explored by several scientists, including Oliver Lodge (1851–1940), though it was not until 1942 that any such detection was achieved. Lodge constructed a very simple receiver, using a short piece of wire attached to a so-called 'coherer', and screening it from the heat and light of the Sun by a blackboard. He had not the slightest chance of success, having no sufficiently sensitive receiver, and furthermore he was plagued by electrical interference from the city of Liverpool. He is now remembered as an advocate of the aether theory of radio propagation and as the first Principal of Birmingham University.

The Sun is at times a very powerful source of radio waves, and became the most promising target for the early investigations in radio astronomy. As it happened, however, the first extraterrestrial radio waves to be detected were not from the Sun but from a totally unexpected source: our galaxy, the Milky Way. Furthermore, the discovery was a byproduct of a commercial investigation into long-distance communications rather than a targeted scientific investigation. The technology of radio communications had developed rapidly in the first decades of the twentieth century, starting with Marconi, who in 1901 famously sent the first radio signal across the Atlantic. Frequencies below 1 MHz (wavelengths longer than 300 metres) were mostly used; we now call these low-frequency, or long-wave, radio. In 1930 the Bell Telephone Laboratories opened a new research station at Holmdel, New Jersey, intending to develop receivers and antennas for radio communication at higher radio frequencies. However, there was little or no idea of what possible difficulties might be encountered, such as interference from sporadic radio bursts from lightning flashes. A newly appointed physicist, Karl Jansky (1905–1950), was given the task of exploring the level of unwanted radio signals which might form a confusing background to a communications channel. He set out to measure this radio background at 20 MHz (wavelength 15 metres), and to locate the source of any interfering signals. Radio receivers were by then becoming very much more sensitive, using vacuum tubes and systems such as the 'superheterodyne' (which is well known to radio amateurs today). Furthermore, at the comparatively short wavelength of 14.5 metres it was possible for the first time to build an antenna with a beam that could be steered around the sky. Jansky's 'merry-go-round' antenna is shown in Figure 41, and a replica can be seen at the National Radio Astronomy Laboratories in the USA.

The length of the elements in Jansky's antenna was determined by the wavelength. The conducting elements were brass pipes, each a quarter of wavelength (3.8 metres) long, supported on a timber frame. The total length of 29 metres was two wavelengths. This array of resonant conductors, with a similar array behind to act as a reflector, was arranged to receive signals from the broadside direction.[1]

Jansky can have had no idea that he was building the world's first radio telescope. By mounting a simple array on a turntable, he was able to locate any source of radio interference within an accuracy of about 10° (the directional accuracy of any such antenna system is limited by its overall size in relation to the wavelength). The patch of sky over which such an antenna is sensitive is referred to as the 'beam'; in Jansky's merry-go-round the beamwidth[2] was 24°. Scanning this beam round the sky at the rate of one rotation every twenty minutes, and continuing for most of a year, Jansky was able to locate any source of radio signals that might interfere with a communication link. To his surprise, a background noise was always

Fig. 41. The merry-go-round antenna at Holmdel, with Karl Jansky in the foreground.

present, from every direction. This signal, in the form of a steady hiss, was much larger than any noise generated in the electronics of his receiver, and he realized that it must originate in the sky. Furthermore, it varied fairly smoothly by a factor of around 2 as the merry-go-round rotated. Even more significantly, the peak and the whole pattern of variation moved smoothly across the sky from day to day, repeating exactly after a year. This is the pattern of behaviour followed by the stars and the Milky Way rather than by the Sun. The peak occurred when the Milky Way was at the centre of the beam. Jansky had discovered radio waves from our galaxy.

As far as Bell Telephone Laboratories was concerned, the important result of Jansky's work was a measurement of the fundamental limitation on the sensitivity of a radio communication link. There was almost no reaction from the astronomical community, even when in May 1933 the front page of the *New York Times* announced 'New radio waves traced to center of the Milky Way'. It took twenty years before the source of the radio waves was understood as radiation from high-energy particles in interstellar space, rather than some new type of radiation from the Milky Way stars themselves. The high-energy particles were part of the cosmic rays (discussed in Chapter 7), and their radiation was due to the motion of these charged particles in the weak magnetic field which permeates the galaxy.

The First Steerable Dish

The next move came from a radio engineer, Grote Reber (1911–2002), who built the first easily recognizable radio telescope. He set out entirely on his own to scan the sky with a narrower telescope beam, intending to locate the source of the Milky Way emission and understand its origin.

Reber was not only a gifted professional engineer; he was an enthusiast of amateur radio who by 1937 had made contact with amateurs in more than sixty countries by radio and, in his own words, was looking for other worlds to conquer. Using only his own resources and enthusiasm, he built in his own back yard a parabolic reflector 9.6 meters in diameter, with a sheet-metal surface accurate enough for a wavelength as short as 9 cm. Providing for such a short wavelength was a bold step from Jansky's work at 14.5 metres; in fact, he found that the strength of the Milky Way signal fell so sharply with decreasing wavelength that his main observations were made at rather longer wavelengths, around 60 cm and 2 metres. His radio telescope, and the map of emission from the Milky Way which he made at 60-cm wavelength in 1944, are shown in Figure 42.

Reber would not have thought of his dish as a radio telescope: the terminology of radio astronomy became current only in the 1950s. Comparisons with modern dishes are interesting: the focal length, the focus supports, the backing structure, and the mounting closely resemble those of later and much larger versions. The mounting at first allowed movement in elevation only, though the whole dish was moved to near Washington DC in 1947 and mounted on a turntable, so becoming an early altazimuth-

(a)

(b)

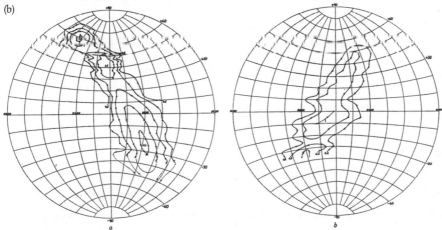

Fig. 42. (a) Grote Reber's 31-foot paraboloidal reflector at Wheaton, Illinois; (b) his maps of radio emission from the Milky Way at 60-cm wavelength.

mounted telescope (see Chapter 3). Reber realized that his design could be scaled up in diameter, and made plans for a 200-foot version. For this he would need institutional support and a considerable sum of money, neither of which was forthcoming. In any case, he preferred to work on his own, and so moved on to other pioneering work at the long-wavelength end of the radio spectrum.

Spin-Off from World War II

Radio astronomy developed immediately and rapidly after World War II. Some of the most talented scientists in the UK and Australia had been drafted into radar development in 1939, and had spent six intense years inventing and exploiting new techniques for radio receivers and antennas. Among them was James Stanley Hey (1909–2000), whose wartime task was to monitor the performance of anti-aircraft radar systems. During the war there was a continuing battle between radar operators and enemy transmitters set up to 'jam' their radars. A particularly serious example, on 12 February 1942, was the jamming of coastal radars which enabled the German battle-cruisers *Scharnhorst* and *Gneisenau* to slip through the English Channel undetected. Hey set out to locate the sources of the jamming, collecting reports from anti-aircraft radars along the coast of southern England. Almost immediately, ten of these radars reported very intense noise-like interference, occurring only in the daytime of 27 and 28 February. This was soon tracked to the Sun; as it happened, at that time there was a huge sunspot, the seat of a very large solar flare.

The discovery of radio from the Sun was one of three major discoveries made by Hey during and immediately after the war. His second achievement was to understand the background of sporadic echoes seen by his radars, which turned out to be from meteor trails in the atmosphere. His third was the most momentous: investigating the Milky Way radio emission discovered by Jansky, he found a discrete and small but very intense radio source which eventually turned out to be the very distant radio galaxy Cygnus A.

Fig. 43. The modified radar antenna used by James S. Hey in the discovery of Cygnus A.

All three of Hey's discoveries were followed up immediately after the war, providing the basis for three major radio observatories: two in England, set up by Martin Ryle (1918–1984) at Cambridge and by Bernard Lovell (1913–2012) at Jodrell Bank, and another in Australia, set up by Joe Pawsey (1908–1962) and John Bolton (1922–1993). Lovell initially set out to use one of Hey's radars to detect echoes from ionized clouds created by cosmic-ray showers (see Chapter 7); but finding none, he turned to an investigation of meteor trails, which became a classic delineation of their origin. Ryle and Pawsey started with the Sun, finding ways to locate the source of radio waves within the 0.5-degree disc. Bolton was the first to concentrate on Cygnus A, though this soon became the overwhelming interest of all the new radio observatories.

Hey's radars were simple transportable systems, operating at a wavelength of 5 metres and using Yagi arrays similar to modern domestic television antennas. In his discovery of Cygnus A, he used a modified antenna consisting of four Yagi aerials mounted on one of the original radar cabins (Figure 43). This simple antenna is hardly recognizable as a radio telescope, and the next developments continued to follow similar patterns which would be more familiar to the radar scientists and engineers than to traditional optical astronomers. Lovell was the exception; he built a large parabolic reflector which became the forerunner of the famous 250-foot Lovell Telescope, which continues in operation to this day. The other groups started a new development, the radio interferometer, which now dominates the whole of modern radio astronomy. But first we will trace the growth of the parabolic reflectors, or 'dishes' as they are popularly called.

The Big Dishes at Jodrell Bank

When Bernard Lovell returned to Manchester University after his radar work in World War II, his main interest was in cosmic-ray showers and the possibility of detecting them by radar. Needing to detect a faint radar echo from high in the atmosphere, he built a large dish which looked only upwards. For a reflector working on a comparatively long wavelength of around 1 or 2 metres, there was no need to make a continuous sheet-metal surface. He could therefore construct very cheaply a large reflecting surface 218 feet across, using only stretched wires. It was erected entirely with local labour, including students and families, at a cost of £2,000, and was first used in 1948. The focus, with simple dipole feeds, was 126 feet above the ground, at the top of a tall mast. Robert Hanbury Brown (1916–2002) equipped this dish with a sensitive receiver for a wavelength of 1.9 metres, and in 1950 achieved a momentous detection of radio emission from the nearest neighbouring galaxy, the Andromeda Nebula. To make a map of this, the first extragalactic radio nebula, the 2-degree beam of the telescope had to be swung out of the vertical by up to 15° by tilting the mast by this large angle—a precarious and laborious manoeuvre involving the adjustment of eighteen steel guy wires. Lovell realized that a dish like the 218-foot could be of immense value if it could be made with a more accurate surface, allowing the use of shorter wavelengths, and crucially if it could be mounted so that it could be steered to anywhere in the sky. So was born the concept of the big steerable dish, which was eventually achieved in the 250-foot Mk I radio telescope (Figures 44a and 44b), now known as the Lovell Telescope.

Lovell found help and inspiration from Charles Husband (1908–1983), a consulting engineer experienced in bridge building. The mounting was the first large altazimuth system, with the dish slung between two towers on a frame which could move round a circular track. Figure 44a shows the telescope as originally built, and Figure 44b after subsequent improvements. The azimuth movement is essentially unchanged today, but the elevation supports are now supplemented by a massive circular arc which carries part of the load and provides a smoother and more accurate elevation drive (the original lightweight arc was added to control oscillations, and was not load-bearing). In October 1957 the telescope was almost complete after a series of financial crises which had left a serious deficit and threats of legal actions. The situation was transformed when the first artificial satellite, Sputnik I, was launched by the USSR. The success story of the telescope starts at that dramatic moment in space exploration.

It emerged immediately after the launch of Sputnik I that there was no operational radar which could track the launch rocket, which was also in orbit. This huge rocket was a prototype for an intercontinental ballistic missile. The defence of Western countries required a radar which could monitor the launch of such vehicles from anywhere in the USSR. Fortunately, Lovell's research team had the solution. By adapting a radar designed for meteor research, and installing it on the nearly completed telescope, they were able to track the rocket, to world-wide acclaim. The funding deficit was finally cleared by a donation from the Nuffield Foundation, and the Mk I 250-foot radio telescope began a research career which has continued for over half a century.

The main cause of the funding shortfall in the Mk I project was a change in the specification of the reflecting surface. The original design, based on experience with the 218-foot dish, specified a wire-mesh surface which

Fig. 44. (a) The original 250-foot Mk I radio telescope at Jodrell Bank, under construction in 1957. (b) The 250-foot radio telescope at Jodrell Bank in its later form, the Mk IA, renamed the Lovell Telescope.

Plate 14. The Major Atmospheric Gamma-Ray Imaging Cherenkov (MAGIC) air-shower telescope. A stereoscopic pair in operation on La Palma in the Canary Islands.

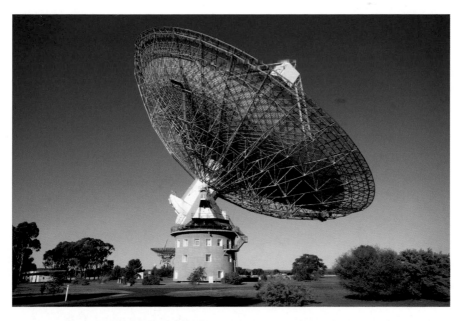

Plate 15. The 210-foot radio telescope at Parkes, Australia. The central part of the reflector surface is sheet metal, and the outer part is mesh.

Plate 16. The Nançay radio telescope. The focus of the parabolic reflector is on the ground. The horizontal beam is reflected to any angle by the flat reflector, which can be tilted to an angle of up to 45°.

Plate 17. The Five Hundred Metre Aperture Spherical Telescope (FAST), under construction.

Plate 18. The Hercules A radio galaxy, imaged by the VLA and superposed on a picture of the visible galaxy from the Hubble Space Telescope. At the centre, a black hole energizes twin jets of energetic particles more than 1 million light-years long, far outside the visible galaxy.

Plate 19. The 10-metre Radioastron antenna folded ready for installation in the launch vehicle.

Plate 20. The Atacama Large Millimetre Array (ALMA) in the Atacama desert. This picture shows sixteen of the sixty-six dishes, concentrated in a compact array at the centre.

Plate 21. The protoplanetary disc surrounding the young star HL Tauri. This image, taken by ALMA at 1 mm wavelength, is sharper than is routinely achieved in visible light with the NASA/ESA Hubble Space Telescope. The whole picture is only 2 arcseconds across. These ALMA observations reveal substructures within the disc which are the possible positions of planets forming in the dark patches.

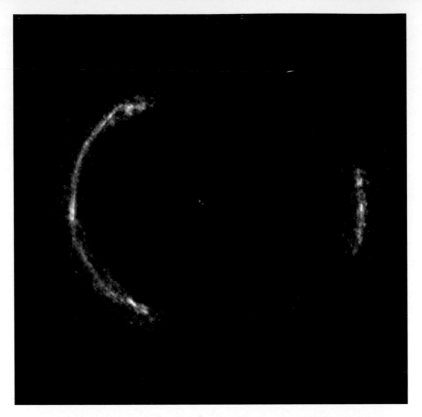

Plate 22. An Einstein ring observed by ALMA. The very distant galaxy seen as a ring is aligned exactly behind an unseen massive galaxy, which acts as a gravitational lens.

Plate 23. The balloon launch of BOOMERANG, which measured the ripples in the Cosmic Microwave Background. The balloon flight, at an altitude of 39 km over Antarctica, continued for ten days.

Plate 24. The Cosmic Background Imager (CBI). Operating at 3-cm wavelength at a dry site in the Atacama desert from 1999 to 2008, this telescope measured the fine structure in the Cosmic Microwave Background.

Plate 25. The background sky measured at microwave radio wavelengths imaged by Planck. The map is oriented with the Milky Way along the equator.

Plate 26. The Square Kilometre Array (SKA), an artist impression showing part of the central concentration of dishes.

would reflect efficiently for wavelengths of 1 metre and longer. However, in 1951 the dramatic discovery of a radio spectral line in Galactic radio emission forced a change in the specification to operation at the much shorter wavelength of 21 cm. The surface now had to be made of sheet steel, at greater cost and with a more massive supporting structure. The reflecting surface was a continuous welded sheet, accurate to within a few centi- metres, the worst deviations from a true paraboloid occurring from gravi- tational bending when the whole telescope was pointing away from the zenith and towards the horizon. In a major rebuild, the new elevation support reduced this problem, and a new segmented surface was installed to allow operation at even shorter wavelengths. The new surface had a shallower profile, so that the focal length was increased and the receiver cabin was raised to its present and more prominent position.

The Mk I, renamed the Lovell Telescope in 1987, had several other adventures in the early Space Age. Its size gave it a unique sensitivity for monitoring radio transmissions from distant spacecraft, such as the Russian Luna 2 which made the first landing on the Moon in 1959. Observations at Jodrell Bank by John Davies tracked Luna 2 as it accelerated towards the Moon, and proved to a hitherto sceptical world that Russian space technol- ogy was well ahead of that of the USA. In 1966, Luna 9 made a soft landing on the Moon, and transmitted pictures of the surface which were recorded by the 250-foot telescope and reached the front page of the *Daily Express* the next day. The telescope also made a vital contribution to the American space programme, when in 1960 NASA's deep space probe Pioneer V depended on Jodrell Bank to command the probe and record its transmitted data. Pioneer V was tracked and its signals picked up out to a distance of 39 million km.

The Big Dish at Parkes

A similar role in space research was played by the 64-metre radio telescope at Parkes, in Australia. This telescope (Plate 15) was the brainchild of another veteran of radar in World War II, Edward (Taffy) Bowen (1912–1991), who achieved the first installation of radar in an aircraft in 1937, and played an important part in the introduction of centimetric radar techniques to the USA. After the war he persuaded the Carnegie Corporation and the Rockefeller Foundation to fund the construction of the first large, fully steerable radio dish in the southern hemisphere. It was completed in 1961, and in 1969 it was used to relay televised pictures of the Moon from Apollo 11. A film *The Dish* made in 2000 is an excellent, though partly fictional, account of this episode. Because of its geographical position, the Parkes telescope continues to be in demand to supplement other tracking stations; for example, in relaying signals from *Curiosity*, the Mars Rover which landed in 2012.

The design of the 64-metre Parkes dish largely overcame the problem of gravitational deformation, and the surface is good for operation at wavelengths as short as 10 cm. Most of its work, however, has been at 21-cm wavelength. This is the characteristic wavelength of radio emission from hydrogen atoms in interstellar space. Surveys of the whole sky at this wavelength are used to map the dynamics of the Milky Way galaxy. The radio emission from hydrogen atoms in any part of the galaxy reveals the amount of hydrogen in the line of sight. The emission is at a very precisely defined radio frequency, so that any line-of-sight velocity shows up as a Doppler shift. The importance of this wavelength band is recognized in international allocations of frequencies for communications, which avoid transmitting in a wide band around the hydrogen-line frequency. The Parkes telescope is especially well equipped for surveys in this wavelength band, including an all-sky survey of hydrogen emission from other galaxies. Surveys are very time-consuming; at Parkes, the process has been accelerated by the installation, at the focus, of an array of thirteen separate receiving systems, each consisting of horns with receivers for two polarizations. Each of the twenty-six receivers is cooled to 20K (degrees absolute) to reduce internal noise. More than half of the presently known pulsars have been discovered in a survey using this system.

The mountings for these huge dishes cannot follow the pattern of most small optical telescopes, which use a polar mount (see Chapter 3) in which the telescope moves about a single axis to compensate for the rotation of the Earth. The required movements for a telescope mounted on a circular track, the altazimuth mounting, must be converted in some way from the celestial coordinates which specify a position in the sky. Digital electronics make light of this task, but the Jodrell Bank and Parkes telescopes were designed before the age of digital computers, and initially had to use electromechanical devices for the coordinate transformation. The Parkes telescope was originally controlled by a scrvo system which locked the telescope onto a small polar-mounted model telescope—a system which nicely illustrates the progression towards modern mountings.

Bigger and Better Dishes

Two more fully steerable radio telescopes, in Germany and the USA, both with diameters of 100 metres, complete the story of increasing size. The first is the Effelsberg 100-metre radio telescope of the Max Planck Institute of Radio Astronomy at Bonn. Completed in 1971, this was for twenty-nine years the largest fully steerable radio telescope. In 2000 it was surpassed by the Robert Byrd Green Bank Radio Telescope (the GBT) at the National Radio Observatory at Green Bank, West Virginia, which had a slightly larger effective collecting area. Although both telescopes have continued to be in great demand since the day of construction, the development of synthesis telescopes, as described in the following chapters, implies that no other such large steerable dishes are likely to be built. Both the Effelsberg and the GBT telescopes, however, incorporate some interesting advances which allow them to be used efficiently at short radio wavelengths.

The Effelsberg telescope was the brainchild of Otto Hachenberg (1911–2001), a German radio engineer who was involved in radar during World War II. By a remarkable coincidence it was Hachenberg who investigated the remains of a British airborne radar set in a crashed Bomber Command Lancaster; this was an H2S radar developed by Bernard Lovell. These two scientists first met in 1977, and were astonished to find that they had both moved from the same wartime background to become pioneer builders of radio telescopes.

Starting some years later than Lovell, Hachenberg realized the importance of shorter radio wavelengths, and determined on a design to remove the effect of gravitational deformations in a steerable telescope. The idea was to allow a deformation to occur, but to arrange that the surface remained as a paraboloid. What would change would be the position of the focus, so that by arranging to move the receiver feed as the elevation of the telescope changed, perfect focusing could be preserved. The detailed design for such a structure was achieved by computer; the result was a surface accuracy specified to be better than 1 mm at all elevations, allowing operation with full efficiency at the short wavelength of 2 cm. A later improved surface reached an accuracy of 0.5 mm; using the centre part of the dish, some observations were made at the even shorter wavelength of 3.5 mm (a frequency of 86,000 MHz).

Another innovation in big dish design was the use of a secondary reflector at the focus, following the Gregorian design of optical telescopes. The secondary reflector had to be large enough to refocus the beam accurately onto a receiver close to the centre of the reflecting surface; for the 100-metre telescope, a mirror 6.5 metres in diameter was needed.

An early use of the Effelsberg telescope was a survey of the sky at the comparably long wavelength of 72 cm, chosen as a wavelength that would show details of the Milky Way and extragalactic nebulae. Since this telescope could not reach the horizon, the 250-foot Lovell Telescope was used to continue the mapping further south, and the Parkes telescope completed the survey into the southern hemisphere. This remarkable collaboration of the three large dishes resulted in a radio map of the sky (Figure 45) which is used to this day.

The last in the line of fully steerable dishes is the Green Bank Telescope (GBT). This has two innovations, both designed to enhance sensitivity through increasing the effective area. The most obvious is the position of the focus, which places the secondary mirror and the receivers completely outside the telescope beam (Figure 46). This means that the main reflector

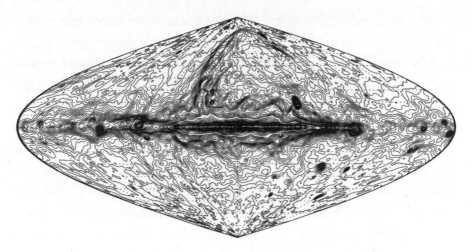

Fig. 45. A map of the Milky Way at 72-cm wavelength (408 MHz).

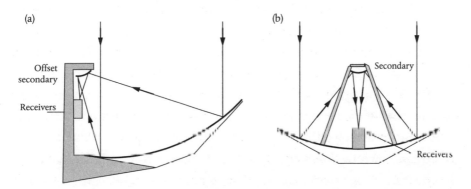

Fig. 46. Avoiding aperture blocking. (a) The offset feed of the Green Bank Telescope. (b) The symmetrical Gregorian focus as in the Effelsberg radio telescope. The secondary mirror and its support in the symmetrical arrangement inevitably obstructs part of the telescope aperture.

surface is not symmetrical; it is a section of a much larger paraboloid which has the secondary mirror at its focus. The aperture is an unobstructed ellipse of 100 × 110 metres. The second innovation is a comprehensive surface adjustment system which moves and tilts each one of 2,004 individual surface panels. The accuracy is sufficient to allow efficient operation at 3-mm wavelength. At such short wavelengths there are many radio spectral lines originating in molecular clouds in the Milky Way and other galaxies, and observations at the GBT have made it a centre for astrochemistry. At longer wavelengths the GBT has made many discoveries in pulsar astronomy—most notably in searching for radio counterparts of pulsars discovered by the gamma-ray observatory Fermi LAT (Chapter 7).

Dishes Looking Upwards

Building even larger fully steerable radio telescopes remains an unfulfilled aspiration for astronomers, but size does bring sensitivity, and other means had to be found to increase the collecting area. The biggest advances have been made in an entirely different direction, aperture synthesis, which is the subject of later chapters. There is still, however, considerable advantage in using a single large dish. Following the early example of the 218-foot fixed dish at Jodrell Bank, in the early 1960s two large parabolic reflectors were built, both completely fixed but with entirely different methods of steering the direction of the telescope beam. The first, at Nançay, in France, is shown in Plate 16. The mechanical engineering involved in this system is much simpler than tilting the paraboloid itself. Basically a transit telescope, the beam can be swung to follow a source by a sideways movement of the receiver at the focus, tracking any source for up to an hour. This telescope has been in continuous operation since 1964.

A different approach is adopted in the much larger fixed reflector at Arecibo, Puerto Rico. This 1,000-foot reflector (Figure 47) has a spherical rather than a parabolic surface, which allows the beam to be swung over a large angle but requires a more complicated arrangement at the focus. The Arecibo radio telescope was originally designed by William E. Gordon (1918–2010) of Cornell University, for research into the terrestrial ionosphere. The intention was to use radar reflections from the Moon and the planets to probe transmission through the ionosphere. Radar is still an important part of the programme; the Arecibo dish has used radar to produce surface maps of planets and some outstandingly accurate measurements of distances in the Solar System. As a sensitive receiving system it

Fig. 47. The Arecibo 1,000-foot radio telescope.

has proved particularly useful for detecting weak radio from individual objects such as pulsars. In this field, the Arecibo dish yielded several major discoveries: the 33-millisecond periodicity of the Crab pulsar, the first binary pulsar, the first millisecond pulsar, and the first pulsar planetary system.[3]

The Arecibo dish was completed in 1963, and has since been considerably improved in several stages. A site for such a large dish is hard to find, as it needed a huge dry depression 50 metres deep extending over 7 hectares (18 acres). Such a natural depression was found in the limestone karst landscape in Puerto Rico. The original surface was a wire mesh, but this is now replaced by 38,778 perforated aluminium panels, sufficiently accurate for operation at the short wavelength of 3 cm. The feed arrangements at the focus provide for sideways movement of massive apparatus including radar transmitters, so that the beam can be swung over an angle of 20°

from the zenith. The spherical surface does not by itself have a simple focal point; in conventional optical terminology it suffers from spherical aberration. Although not the whole of the 305-metre aperture is used, but only an area depending on the direction of the telescope beam, this is a very serious effect, particularly at shorter wavelengths. The most remarkable and spectacular remedy to this problem can be seen prominently in Figure 47: a dome 30 metres across, facing down into the main dish, containing two mirrors which refocus the radio wave. In traditional terminology this is a Gregorian system, with the mirror surfaces shaped to compensate for spherical aberration. From an engineering point of view it is an unusual structure. Suspending a 900-tonne dome 150 metres in the air is an achievement in itself, but this structure has to be kept in place with millimetre accuracy. Furthermore, the whole receiver system has to move sideways by up to 50 metres to provide the maximum swing in beam direction.

The Biggest Dish

For over half a century the Arecibo dish was the largest in the world. Its continuing productivity, coupled with recent improvements such as the use of multiple feeds and wideband receivers, indicated that an even larger dish would be well worth building, provided that a site could be found and new ideas developed to overcome the problems inherent in a spherical reflector. This has been achieved in the Five Hundred Metre Aperture Spherical Telescope (FAST) in Guizhou Province, south-west China. As at Arecibo, the telescope fills a deep depression in an extensive karst limestone formation, fortunately situated in a sparsely populated area in which radio and other electrical interference is minimal.

The dramatic increase in size over the Arecibo dish, and the need for multiple feeds, while preserving the capability of swinging the beam over a large angle, requires a radical change of design. The basic idea of a spherical bowl, with a feed that can move sideways to swing the beam, will be the same, but instead of a fixed spherical reflector surface, with an elaborate system of correcting spherical aberration at the focus, the correction will be achieved at the surface, which will be flexible. At any time, an area only 300 metres across, of the overall 500-metre diameter, will be in use. This area will be distorted into a paraboloid, removing the aberration and providing a simple focal point. This requires the surface to be adjusted by no more than 1 metre, but of course it must be continually adjusted as the telescope beam is moved across the sky. The surface is therefore constructed from 4,600 triangular aluminium panels, curved to conform with the spherical bowl and connected with flexible joints and a network of steel cables. At 2,300 points the supporting network is pulled from behind by cables attached to the ground through winches. Moving the beam then requires a continuous adjustment of 2,300 electric motors behind the dish, as well as the sideways movement of the feed with the receiver systems. The overall accuracy of the telescope, including this active surface and the positioning of the feed system, will be within 5 mm, allowing the beam to be pointed to an accuracy of 4 arcseconds.

The receiver systems of FAST will be in a feed cabin suspended 140 metres above the surface and moveable sideways by up to 110 metres to swing the beam by 40° from the zenith. The cabin is suspended by cables from six 100-metre high towers round the periphery of the dish, with winches to move the cabin over 110 metres along a circular arc, as shown in Plate 17. The cables will adjust the position to around 10 cm, and a fine movement system will then adjust the tilt angle and the position to 5 mm. This is altogether a most remarkable assembly of mechanical engineering.[4]

FAST will operate initially at frequencies between 300 and 3000 MHz. Modern receiver techniques allow observations simultaneously over a wide frequency band and also with multiple beams. The initial instrumentation will be for a wide bandwidth centred on the 21-cm hydrogen line, using nineteen beams, each with two polarizations, feeding a total of thirty-eight cooled low-noise receivers. This system will allow a dramatic extension of several branches of astronomy, mainly in what may be termed the Gaseous Universe. In our Milky Way galaxy, this means concentrating on the structure of the interstellar medium, including the gas clouds which extend far above and below the plane of the galaxy. The frequency range will include spectral lines from hydrogen, hydroxyl OH, and formaldehyde H_2CO. Galaxies in the Local Universe will be accessible up to a moderate redshift, with possible new studies of their populations of pulsars and the dynamics of the gas clouds.

It takes only a brief look at the sheer size and complexity of the 500-metre FAST to realize that there will probably never be another such dish built anywhere in the world. This does not, however, mean that the continued increase in size and sensitivity of radio telescopes is coming to a halt. A new generation of telescopes with even larger apertures is already becoming available. These are all based on aperture synthesis—a concept which derives from the radio interferometers described in the next chapter.

9

PAIRS AND ARRAYS

Michelson's Interferometer

It probably never occurred to Galileo that a telescope might one day make an image of a star as a disc rather than a point of light blurred by diffraction and scintillation. The resolution of his telescopes was not much better than 1 arcminute. Large optical telescopes now achieve better than 1 arcsecond, and with active optics they may reach twenty times better, at 50 milliarcseconds. This is close to the resolution needed for measuring the angular diameters of the nearest large stars, such as the familiar bright star Betelgeuse in the constellation of Orion. Better resolution can be achieved only with larger apertures, since the limit is simply determined by the ratio of the diameter of the aperture to the wavelength of light. In 1920, Albert Michelson (1852–1931) showed how this limit could be overcome by extending the aperture, combining light from two separate mirrors to make a single image. This was the beginning of a new era in telescope design. As it happened, the principle was realized in radio astronomy more than in optical astronomy, eventually giving us the extensive arrays of radio telescopes that now dominate the scene, leading up to the huge international Square Kilometre Array.

A telescope that combines light or radio waves from two apertures to make a single image is called an interferometer, and the multi-element radio arrays which effectively build up a large single aperture are called synthesis arrays. Michelson's interferometer is shown in Figure 48; the two apertures were mirrors mounted on a beam across the front of the 100-inch telescope on Mount Wilson in California. Light from these mirrors was fed

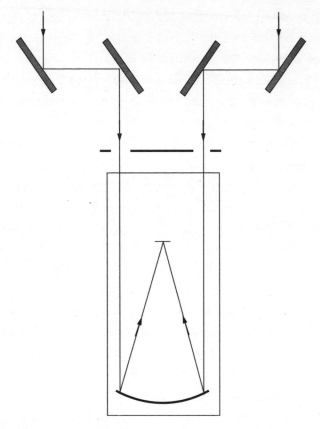

Fig. 48. Michelson's stellar interferometer, showing how light from two separate apertures was fed into the 100-inch telescope.

into the telescope with another pair of mirrors, and an image was formed as usual at the telescope focus. The effect of the two separated apertures was seen within the image, which was crossed by a pattern of ridges known as interference fringes.[1] Only a small point-like star would show these fringes; in an image of a star with a larger diameter the fringes would be smoothed out. Betelgeuse showed this effect of smoothed-out fringes. In 1920, Michelson and his colleague Francis Pease had shown that this star had an angular diameter of 0.05 arcsecond. No comparable measurement of the size of a star was made for more than another half a century.

Michelson understood that a simple interferometer could in theory be duplicated many times with different spacings between the two apertures, producing essential information for building an image of the disc of the star. In 1907 he was awarded a Nobel Prize for his measurement of the velocity of light, and he is also well known for another fundamental experiment using an interferometer, the Michelson–Morley experiment, which disproved the existence of an aether in which light propagates.

The next realization of the interferometer principle was in the radio rather than in the optical domain. The reason for this is simply that the radio wavelengths used in early radio astronomy were a million times greater, so that constructing apparatus with an accuracy related to wavelength was very much easier. During World War II, James Hey had discovered that the Sun emits intense radio waves at the time of a solar flare (Chapter 8), but no-one had detected radio from the quiet Sun. The challenge was to detect this against the bright radio background of emission from the Milky Way galaxy, which extended over the whole sky. A radio telescope was needed that could distinguish the small source of the Sun from this intense background, but no radio telescope existed with sufficient angular resolution. The problem was solved with the invention of the radio interferometer.

The angular diameter of the visible Sun is 0.5°—many orders of magnitude larger than the size of Betelgeuse. A telescope aperture would have to be more than a hundred wavelengths across if it was to have a beam narrow enough to pick out the Sun, and much larger if it was to have a beam narrow enough to make a map of the sources of radio on the disc of the Sun. During World War II, two groups of scientists working on radar had become familiar with the techniques needed for a radio interferometer, and both used interferometers to achieve the detection of the quiet radio Sun soon after the war. The closest analogy to Michelson's interferometer was built in Cambridge (England) by Martin Ryle (1918–1984).[2] Using wavelengths of around 1 metre, he built two simple antennas, spaced them several wavelengths apart, and connected them to a single receiver (Figure 49a). The quiet Sun was easily detected, but while he was working the Sun became very active and produced a huge radio outburst. Did this come from the whole Sun, or only a small part of it, probably above a

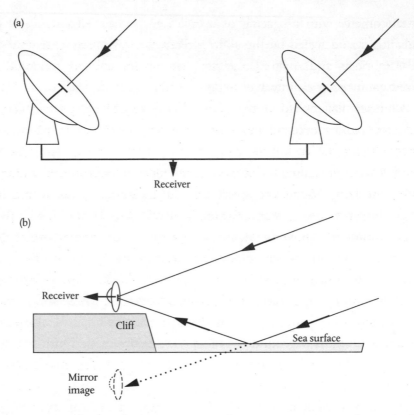

Fig. 49. (a) A radio interferometer. A pair of antennas are connected to a single receiver. (b) Pawsey's clifftop interferometer used a single antenna, which became an interferometer when combined with its reflection in the sea.

sunspot? To resolve this question a much better angular resolution was needed, so Ryle moved the two antennas further and further apart until he could show that the source was indeed very small. Ryle recognized that his interferometer was basically the same as Michelson's, and later named a larger version after him.

The other radar pioneer to use an interferometer was Joseph Pawsey (1908–1962). Radar in Australia during World War II was developed to detect aircraft flying in from the sea, and radars were installed on top of high cliffs, looking towards the horizon. Inevitably the radars picked up two signals from the target—one direct and the other reflected from the sea (Figure 49b). The radar antenna and its reflection in the sea formed an interferometer with a spacing of over a hundred wavelengths. Sunspot radiation was detected by the clifftop receiver in the same year (1946) as in Cambridge, and Australian radio astronomers later concentrated on investigating solar radio waves in detail.

A further challenge for both groups was presented by Hey's discovery of a discrete radio source hidden in the Milky Way background. This became known as Cygnus A, but its nature was completely unknown. Ryle and his colleagues (including the author) then built an interferometer, which he called the Long Michelson, specifically for locating Cygnus A and any other similar, as yet mysterious, radio sources. A spectacular result from this experiment is shown in Figure 50. Here the signal recorded in a single day shows two prominent radio sources: Cygnus A, and another new source now known as Cassiopeia A. The Galactic radio background varies smoothly through the whole record, and only the point-like sources show the interferometer pattern. Locating these sources was a very exciting task. Accurate positions were found by the Australian group for another source, Taurus A, which was identified with the Crab Nebula. Finding positions accurate to about 1 arcminute for the other sources became my PhD task. With the help of two astronomers at the Mount Wilson Observatory, Rudolf Minkowski and Walter Baade, Cassiopeia A was identified with the remains of a supernova explosion, while in complete contrast, Cygnus A was identified with a very distant galaxy, far outside the Milky Way.

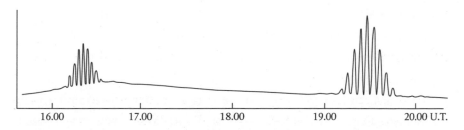

Fig. 50. A recording from the Long Michelson interferometer showing the discrete radio sources Cygnus A and Cassiopeia A on the background of radio from the Milky Way galaxy.

Analysis and Synthesis

What size were these newly discovered radio sources? Such an intense source of radio energy surely could not be as small as a visible star. The radio version of Michelson's interferometer should provide the answer,[3] but at such long wavelengths it involved very large spacings between the two antennas, perhaps many kilometres. Three groups, at Cambridge and Jodrell Bank in England and at Sydney in Australia, set out to construct interferometer pairs with large spacings. At Cambridge we used all the coaxial cable we could find to connect the pair, while the others developed a way of connecting their interferometer pairs by a radio link. We at Cambridge were lucky; the cable link was simpler, and the spacing was just sufficient to show that both sources were several arcminutes across.[4] The Jodrell Bank observers went further; they measured the visibility of the interference fringes at larger spacings, and showed that this gave the structure of the radio source as well as simply its diameter. This was the analysis which Michelson had foreseen. The Jodrell Bank analysis showed that the Cygnus A radio source was double, with two separate clouds of radio emission far outside the visible galaxy.

Soon, larger interferometric radio telescopes, built at Cambridge and Sydney, were finding many more radio sources, most of which had no obvious visible counterparts. The Jodrell Bank radio astronomers set out to measure the diameters of these new sources, and found that some of them were so small that very large interferometer spacings were needed. The radio links were extended so far that relay links were needed. The largest spacing used—134 km—extended across the country to Yorkshire. Surprisingly, a small number of radio sources were still unresolved. They seemed to be like stars rather than the extended clouds of the first known radio sources, and they were dubbed Quasi Stellar Radio sources, or Quasars, a now familiar name. Quasars are the rare, very energetic centres of some very distant galaxies, whose energy is derived from stars and gas clouds falling into a massive black hole.

Analysis of the size and structure of a radio source was the first step in a momentous move towards the concept of aperture synthesis, in which separate antennas are combined to form a single, much larger, radio telescope. Could observations using many different interferometer spacings be put together to give an actual two-dimensional image of the source, like the image that would be produced using a huge telescope with a diameter equal to the largest interferometer spacing? The theoretical possibility was clearly understood by all three research groups, but there were two practical problems to be overcome. The first was to make a stable connection of the pairs of antennas so that the interference pattern could be recorded with both phase and amplitude. The second was the large computation needed to combine the recordings to produce a reasonably accurate image; this was to be achieved in an era before the days of digital computers.

The basic idea of this process of aperture synthesis is shown in Figure 51. A pair of small portable antennas observes the same object at a sequence of spacings which eventually covers the whole large area. Adding the records from all observations gives the same result as would be produced in a single observation made with the full aperture. This tedious process was in fact used for the first demonstrations of aperture synthesis, in which a map of radio emission from the quiet Sun was made at Cambridge.[5] A far more efficient process is shown in Figure 51b. Here the individual small antennas only need to be placed at successive stations on an east–west line, observing continuously as the Earth rotates.[6] In the first demonstration of this Earth rotation synthesis, the individual antennas were segments of a long array already built for a previous survey, and the target area of sky was centred on the North Pole. The result of adding many days of recordings was the spectacular map shown in Figure 52.[7]

Covering an area of sky 8° wide, centred on the North Pole, this map contained fifty-three discrete sources, mapped with a precision of 4.5 arcminutes. At the long wavelength of 1.7 metres, this was the equivalent of the performance of a single dish more than 1 km in diameter. The map was a revelation, not only of the unforeseen huge population of individual radio

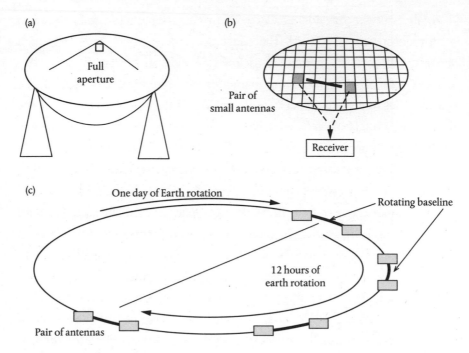

Fig. 51. Aperture synthesis. A large radio telescope (a) is to be built up from sequential observations using only a pair of small antennas placed at all possible spacings within the desired large aperture. (b) More efficiently, a recording using a single pair and continued for twelve hours covers a full range of angles. (c) This covers an elliptical track as the Earth rotates, so that the pairs of antennas need only be placed at a series of spacings on an east–west line.

sources in the sky, but also of the possibility of building much larger radio telescopes than could be constructed as traditional single dishes. The 1974 Nobel Prize for Physics was awarded to Martin Ryle for this invention, along with Antony Hewish for the discovery of pulsars.

The Cambridge One Mile Telescope

The next stage in aperture synthesis was to focus on an area of sky away from the North Pole, which involved telescope elements which could track any region of sky rather than stay fixed on the pole. In 1964, Martin Ryle built the first telescope specifically designed for Earth rotation synthesis,

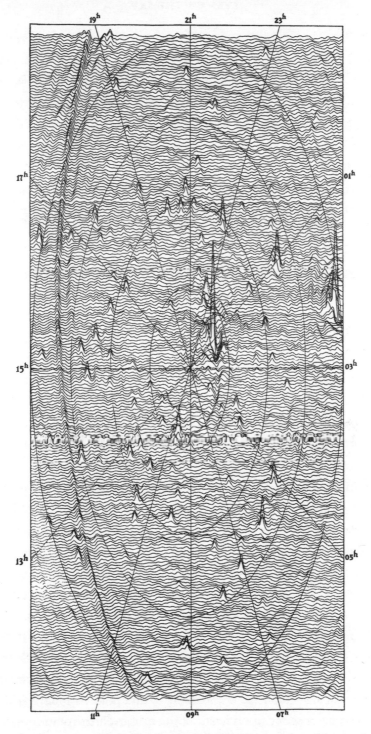

Fig. 52. The first map of the radio sky made by Earth rotation synthesis. The map, centred on the North Pole, contains fifty-three discrete sources.

with the baseline extending to 1 mile and using three polar-mounted dishes, 18 metres in diameter. The baseline was again aligned east–west. One dish was mobile, on a railtrack half a mile long, while the two others were fixed so that all spacings up to 1 mile were available. Using wavelengths of 75 and 21 cm, this telescope mapped selected regions of sky with a resolution of 20 arcseconds, and was the first to produce radio images with a resolution better than the human eye.

The principles behind aperture synthesis had been known since the early 1950s. The relation between the output of a radio interferometer and the complete image of an object was first set out by Australian radio astronomers as early as 1947.[8] Mathematicians will recognize that this concept is best expressed in terms of Fourier analysis. The practical problem was to construct a full image from all the interferometer outputs—a process requiring a computer with a capacity that did not exist until electronic computers became available. The role of the digital computer in allowing aperture synthesis to develop is analogous to the role of detector arrays in allowing optical telescopes to map huge areas of sky in a single observation. In Cambridge it was the series of computers set up by Maurice Wilkes—Edsac I (1949–1957), Edsac II (1957–1965), and Titan (1963–1973), each the most powerful available at the time—that enabled the successive steps in aperture synthesis. Titan was the first to use transistors rather than thermionic valves. All the recorded data, and the specially developed programmes, were written on punched paper tape. The development of computers was so fast that the new telescopes could be designed on the assumption that sufficient computing capacity would be available by the time construction of the telescope was complete. (This still happens today. The Square Kilometre Array is so demanding of computer capacity that the design is dependent on computers that have yet to be designed and built; see Chapter 12.)

The One Mile Telescope was a resounding success. The first maps it produced showed sources a hundred times fainter than had been seen before. Individual sources, such as the supernova remnant Cassiopeia A,[9] showed structure with detail comparable to optical photographs (Figure 53), and it became possible to understand the extraordinary physical processes

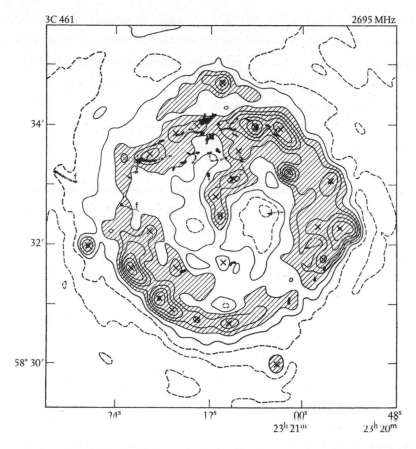

Fig. 53. The supernova remnant Cassiopeia A, mapped in 1970 by the One Mile Telescope at 11 cm wavelength. The contours are radio emission, and the shaded areas show visible emission.

involved in supernovae, radio galaxies, and quasars. Cosmology took a great leap forward, with counts of the faint extragalactic sources reaching back in time to show an evolving Universe. Design for the next step started immediately, and in 1971 the Five Kilometre Array, soon renamed the Ryle telescope, went into operation. This Earth rotation synthesis telescope used eight equatorially mounted 13-metre dishes, again spaced on an east–west line. Four of the dishes were mobile, on a rail-track 1.2 km long. Using the shorter wavelength of 2 cm, this improved the angular resolution to 1 arcsecond. All possible pairs of antennas in the array could now be

165

connected as interferometer pairs, and their signals processed in the increasingly powerful computers which were becoming available. This telescope was reconfigured in 2004, to form a compact array for cosmological work (described in Chapter 11).

The Very Large Array

Telescope arrays mounted on an east–west line rely on the rotation of the Earth, and continuous observations for a full twelve hours are necessary to map a single region of the sky. Furthermore, without a north–south component they cannot operate at all near the celestial equator. A solution to this problem was devised by another pioneer of radio astronomy, Bernard Mills (1920–2011) in Australia, who in 1954 built an interferometer consisting of two long narrow antenna arrays, one extended east–west and the other north–south. The pair formed a cross, like a plus sign. A later version of this Mills Cross consisted of two cylindrical paraboloids, each 1,600 metres long. The east–west component is still in use.

The practical problems of the east–west synthesis arrays were overcome in the next dramatic step forward, the Very Large Array (VLA), built in New Mexico from 1973 onwards and completed in 1980 (Figure 54). The VLA is still in full operation (in 2015). With a substantial upgrade in 2011 it was renamed the Karl Jansky Array, but it is still usually known by the acronym VLA. Originally it had twenty-seven dishes, now increased to thirty-six, all 25 metres in diameter, arranged in a Y formation with the three arms each 21 km long. All combinations of antennas can be used, producing interferometer spacings of up to 36 km. Wavelengths from 4 metres to 6 mm (frequencies 73 MHz to 50,000 MHz) are available, spread over sets of eight receivers at each dish. Connections between the dishes must be very stable, so that the signal paths are accurate to a fraction of the shortest wavelength. This was originally achieved with a system of waveguides, and more recently by fibre optics. The angular resolution from the longest to the shortest wavelengths ranges from 25

Fig. 54. The Karl Jansky Array, formerly the Very Large Array, and still known as the VLA. The thirty-six dishes are arranged in a Y formation. They are mobile, and the array can be extended to produce baselines of 36 km.

arcseconds down to 0.04 arcsecond, providing images directly comparable with the best optical images (Plate 18).

With so many individual antennas all in operation simultaneously, there are sufficient baselines between pairs of antennas for an image to be synthesized without using Earth rotation. Such observations are known as the 'snapshot mode', which may be achieved in a short integration time of only ten minutes, enabling many more observations to be made in a single day. The full sensitivity still requires a twelve-hour observation, producing a map which shows thousands of individual radio sources. A single map from the VLA of one area only 0.5° across (defined by the beamwidth of the individual dishes) contained sufficient new and very distant radio sources for the evolution of the Universe to be clearly demonstrated.

Although for many observations the VLA performs like a single huge telescope 36 km across, there is an obvious difference: only a small proportion of the full area is actually receiving the radio signal. This particularly affects observations for extended sources such as gas clouds in the Milky Way galaxy. For such observations the performance of the VLA can be greatly improved by reducing the spacing between the dishes by moving them closer together. Each dish, weighing more than 200 tons, is mounted on a rail-track and can be moved to a compact configuration, reducing the maximum spacing from 36 km to only 1 km.

Longer and Longer Baselines

The discovery that quasars contain radio sources so compact that they could not be resolved by long-baseline interferometers, linked first by cable and then by radio, was taken as a challenge by several groups of engineering-minded radio astronomers. Astronomers who were concerned with very accurate measurements of positions of stars and galaxies also became interested; if it might be possible to relate the positions of stars in the Milky Way galaxy to those of very distant quasars, there might be a fundamental improvement in the reference frame within which all astronomical positions are specified. Extending the baselines of arrays such as the VLA, which was designed to make complete maps, seemed to be impossible.

A more limited objective was to measure positions and diameters of quasars, using only pairs of antennas as interferometers, but at much larger separations. The practical problems, however, were formidable. Extending cable or radio links to connect antennas hundreds or thousands of kilometres apart had to be abandoned; the practical solution was to record the radio signals at each antenna, and bring the recordings together after the event. A recording system with a very high capacity was needed for hour-long recordings of a broad-bandwidth signal. As in many advances in astronomy, and especially in radio telescopes, the solution was found from the rapidly growing world of communications and digital technology. The recording system used for television pictures, the video tape, could carry enough recorded signal; the problem then was to play back the tapes from two separate antennas in perfect synchronism. In the first demonstration of such a Very Long Baseline Interferometer (VLBI), by a Canadian Group in 1967,[10] synchronization was achieved by recording a very accurate clock alongside the radio signal. The first results were amazing: there were quasars with angular diameters less than 0.01 arcsecond. With this new technique, pairs of radio telescopes at any distance apart, even on different continents, could become interferometers, with angular accuracies measured in milliarcseconds or less.

Making the Connection

Video tapes and atomic clocks were soon adopted for the rapidly expanding VLBI networks. Was it possible to go further than a simple measurement of diameter? Could the meagre information produced by a simple interferometer pair, or better by several pairs making a simple array, be used to make simple maps? This would have to be done without trying to fill in the whole aperture, as in the original concept of aperture synthesis shown in Figure 51. The telescope built up in this way might have an aperture hundreds of kilometres across, but most of it would be empty and unused. Looked at from the point of view of an optical telescope maker, the result might be like blocking off almost all of the telescope's aperture, and opening only a few tiny holes or short arcs. The images would be a mess—possibly a hopeless mess. Undaunted, radio astronomers started to use such 'unfilled apertures', and showed that useful maps could indeed be made using only a small number of antennas, widely dispersed. The key was to use Earth rotation, as demonstrated in the One Mile Telescope, observing for long periods as the Earth rotated so as to cover long arcs rather than tiny patches in the synthetic aperture. The essential requirement was that all pairs should be interconnected with stable links, either directly or using wide bandwidth recorders. In this context, stability means that pairs of signals must be synchronized within a small fraction of the period of the radio signal, meaning that the electrical length of each link must be controlled to well within a wavelength. The links might be tens or hundreds of kilometres long, and the useful wavelengths were becoming shorter.

In the first such long-baseline array, which comprised six dish antennas spread over distances up to 217 km, the links were made by radio—hence the name, Multi Element Radio Linked Interferometer Network (MERLIN). This array (Figure 55), which is still in operation, is based at Jodrell Bank, using the Lovell Telescope. It is spread over a large part of central England, where most of the tens of millions of inhabitants are unaware that they live within the aperture of one of the world's largest radio telescopes. The longest baseline extends to Cambridge, where one dish of the One Mile Telescope was used initially, and was replaced by a new 32-metre dish in 1990. At a wavelength of 6 cm the angular resolution is 40 milliarcseconds, which equals that of the Hubble Space Telescope. Interconnections were originally achieved by using microwave radio links, which had been

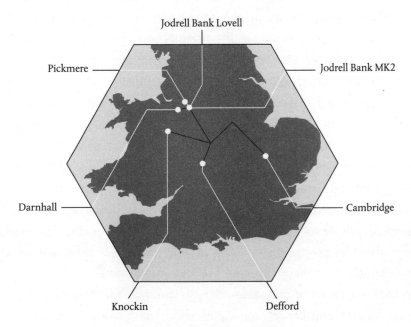

Fig. 55. The antennas of the MERLIN array.

developed for a national telephone network. All six dishes are now connected through fibre-optic cables.

Aperture synthesis on such a large scale proved to be so useful that several long-baseline arrays were established on a continental and eventually an intercontinental scale. At first in the USA, then in Europe and in Australia, arrays of ten or more radio telescopes were combined into networks which produced amazingly detailed maps of quasars and other objects which had seemed impossible to resolve. Combining the data which arrived at a specially built correlator was a major task, not least because of the sheer bulk of the video tapes which had to be transported across the world. In Europe, the correlator is located in the Joint Institute for VLBI in Europe (JIVE), at Dwingeloo in The Netherlands. This European network, the EVN, is a remarkable international collaboration. It uses existing large radio telescopes, spread originally across five countries in Europe and now extended to Russia and South Africa. A total of fourteen major institutes are now members. With so many members, and using some of the largest individual radio telescopes in the world, the EVN is the most sensitive telescope available for high-resolution radio astronomy.

The operation of EVN requires essentially no special funds for the individual telescopes, which are already funded in their own countries. The correlator at JIVE is supported by European Union funds. This arrangement involves the minimum of bureaucracy; the collaboration is purely scientific, organized and run by the participating astronomers themselves. Arranging an observing programme may require more than a dozen telescopes to operate simultaneously, interrupting their own programmes for several days. Furthermore, the results of an observation are often attributed simply to the one observatory that proposed the observation, rather than adding all the names of the individual observatories that participated.

A major network of radio telescopes in the USA, the Very Long Baseline Array (VLBA) consists of ten 25-metre dishes, constructed solely for VLBI. These are spread over a distance of 8,500 km. The shortest wavelength of 3 mm gives an angular resolution of below one milliarcsecond. The ability of the VLBA to measure angles to such great accuracy provides opportunities both for establishing the basic framework of astronomical positions, known as the reference frame, and also for measuring the movements of stars within that frame. Some pulsars, for example, have been found to be moving across the background of stars with velocities of several hundred kilometres per second. The observations incidentally provide the relative positions of the ten dishes to millimetre accuracy, together with the relation of the baselines to the rotation axis of the Earth. Changes in both these can be observed within a year, due to movements within the crust of the Earth and precession of the rotation axis.

Connecting these arrays over such large distances with links accurate to 1 mm or less is practically impossible. The analysis system has instead to correct for inaccurate or missing information on the timing within the interferometer patterns of each pair. If, for example, there is a truly point source within the field of view, it will only be mapped correctly if the timings, or phases, of the patterns are correct. The analysis can then adjust the timings to produce a perfect point source, so correcting the whole observation. The demands for accuracy of the connections and of the analysis have increased as wavelengths have decreased, and as the radio receivers at the individual telescopes have improved. Receivers now use improved amplifiers with low inherent noise and much larger bandwidths, extending to 1,000 MHz (1 GHz) or more. Another major improvement has been the advent of glass fibre-optic cables, which can be used to connect radio telescopes over distances of thousands of kilometres. They provide very stable broadband links and are used to connect dedicated arrays such as MERLIN, and existing scientific fibre-optic networks can be used for connections over the whole of Europe and beyond. MERLIN and the EVN are often used as an extended and very powerful European array. There is no limit to the possibilities of such interconnections, which can extend to any combination of telescopes on a world-wide scale. As many as twenty radio telescopes can now be used simultaneously in such a Global VLBI network.

Into Space with VLBI

The maximum baseline between a pair of existing radio telescopes on the surface of planet Earth is now limited only by the diameter of the Earth itself, and as that distance is approached they become useless as an interferometer since they cannot simultaneously see the same piece of sky. The only way to use even longer baselines is to mount one of the telescopes on a spacecraft and shoot it into a deep orbit round the Earth. Despite the obvious technical difficulties, this has been achieved with the Highly Advanced Laboratory for Communications and Astronomy (HALCA), launched by Japanese astronomers in 1997, and in Radioastron, launched in Russia in 2011. In both, the telescope was small by terrestrial standards, with diameter 8 metres for HALCA and 10 metres for Radioastron, but by spacecraft standards so large that they had to be unfolded like an umbrella after reaching orbit. Plate 19 shows the 10-metre Radioastron antenna folded ready for installation in the launch vehicle.

Although these are comparatively small diameters, they are used with large ground-based telescopes such as the 100-metre Effelsberg and the GBT, and even with the 25-metre telescopes of the VLBA to make interferometer pairs and arrays. With these partnerships, HALCA and Radioastron have observed several of the brightest and smallest radio sources, such as the central objects in quasars. The baselines are many times larger than the diameter of the Earth. HALCA was launched into an elliptical orbit with a maximum distance 21,400 km from Earth, and Radioastron into a very deep orbit which takes it to 390,000 km from Earth—more than the average distance of the Moon. The angular resolution achieved by Radioastron is measured in microarcseconds.

Considering the efforts required in terrestrial interferometers to interconnect them with cable links stable to an accuracy of around 1 mm, it seems impossible to link a spacecraft moving with a changing velocity of more than 1 km per second with a stationary partner on Earth. However, even in a terrestrial pair the two telescopes are moving due to the rotation

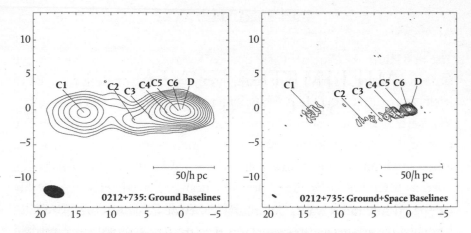

Fig. 56. The structure of the quasar 0212+735 revealed by interferometry with the spacecraft HALCA. The angular scale is in milliarcseconds. The elongated image produced by ground-based observations (left) is seen by including the long spacecraft baselines (right) to be a line of emitters C in a jet from the bright centre of the quasar (D).

of the Earth, and the receivers and analysis systems are already set up to compensate for this. Orbital motion is very smooth, even if rapid, and it can be tracked and followed during the observations.

HALCA achieved a remarkable survey of bright extragalactic radio sources at 6 cm wavelength, using several ground-based telescopes to form an array. Figure 56 shows the dramatic improvement in angular resolution achieved by combining the long spacecraft baselines with the existing ground-based interferometers.[11] The angular resolution of the map is 0.2 milliarcseconds, and the components of this quasar are comparable in size to the distance of the nearest stars in the Milky Way galaxy.

HALCA operated until 2003. Radioastron is still in full operation (in 2015), and proposals are invited for observations at a range of wavelengths from 13 mm to 92 cm.

10

MILLIMETRE WAVES AND SPECTRAL LINES

The cosmic radio waves described in previous chapters are like white light, covering a wide range of wavelengths. These radio waves that tell us so much about the cosmos are generated by energetic particles, usually electrons. The radiating electrons are often in a cloud of hot gas, as for example in the solar corona, and the intensity of their radio emission can then be used as a measurement of the temperature of the gas. The very energetic electrons of the cosmic rays which pervade our Milky Way galaxy also radiate radio waves, as discovered by Jansky in 1935. These similarly radiate with an intensity which depends on their individual energies. In both these sources the radiation is spread over a wide spectrum. There are, however, cosmic radio sources whose radiation is concentrated in a narrow band of frequencies, as in the more familiar spectral lines of optical astronomy, or the sodium light from a street lamp. The light from a star contains spectral lines which are a rich source of information about the star and about its atmosphere, each spectral line being related to its own species of atom or molecule. Similarly, radio spectral lines are related to molecular species, and to one atomic species in particular: hydrogen. The intensity of a radio spectral line is related to both the abundance and the temperature of its related species, which might be a cold cloud of neutral hydrogen or molecules in a nebula energized by a hot star. Some radio spectral lines are, however, spectacularly bright, and we will discuss these so-called 'masers' later in this chapter.

There are spectral lines from hydrogen, and from helium and carbon, which originate in a similar way to optical or X-ray spectral lines, in a step in energy of an electron orbiting round the atomic nucleus. The electron may originally be completely detached, falling in towards the nucleus in a series of steps, and for this reason the process is known as recombination radiation. Atomic hydrogen also has a very special and dominant spectral line due to an interaction between the nucleus and its single electron, known as spin coupling.

The Hydrogen and Hydroxyl Lines

Picking out a spectral line from the Galactic radio emission is like tuning an ordinary radio receiver to find a faint broadcast signal hidden in a noisy background: it helps to know the frequency one is looking for. The discovery of the spectral line from atomic hydrogen became possible when in 1944 the frequency was calculated by Hendrik van de Hulst (1918–2000). The frequency is 1,420 MHz, and the wavelength is 21 cm. As in all spectral lines, the energy of a hydrogen atom can take only one of a series of values, and it is only when the energy switches between these values that the atom radiates. The frequency, or wavelength, of the radiation depends on the energy difference. In hydrogen, the two lowest energy values are very close together, and the radiation is at a correspondingly low frequency, in the radio spectrum.

The frequency of the hydrogen line is very precisely determined by the structure of the hydrogen atom, but any hydrogen line that we observe from the Milky Way galaxy is not a single narrow line. As observed (Figure 57), a typical line is spread over a range of frequencies. The radiation we receive from a cloud depends on its velocity; this is the Doppler effect, familiar in the sound of a siren sounding on a moving vehicle. The spread of frequencies in Figure 57 is directly related to the structure and rotation of the galaxy; the hydrogen gas may be moving towards or away from us at speeds of up to some hundreds of kilometres per second.

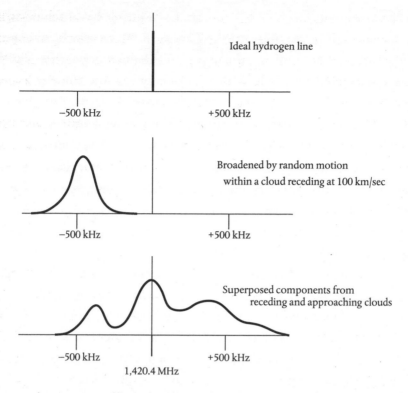

Fig. 57. The structure of hydrogen-line radiation. A spectrum from a line of sight in the Milky Way. The components of this spectrum originate in gas clouds moving with different speeds, which are therefore Doppler shifted by different amounts.

There is a variant form of hydrogen which occurs as a small proportion of the interstellar gas. This is the isotope deuterium, in which the nucleus contains an extra neutron. A deuterium atom has a smaller split between its lowest energy levels, and consequently it has a characteristic spectral line at a lower frequency than a hydrogen atom. This line, at 327 rather than 1,420 MHz, is weak and hard to detect. Haystack Observatory in USA built a large dipole array for its detection, and succeeded in showing that there are only twenty deuterium atoms for every million hydrogen atoms in the interstellar gas of the Milky Way.[1]

Quantum physics, which describes and governs the structure of atoms, also tells us about the strange processes which relate atoms to radiation.

The hydrogen and deuterium lines, at the lower end of the spectrum, provide an extreme example. A hydrogen atom, left to itself in free space, will spontaneously emit one quantum of radiation, a photon, after an average interval of hundreds of thousands of years. The photon it emits can be described as a very long packet of waves, with wavelength many orders of magnitude greater than the size of the atom. How can the atom decide to radiate a photon after such a long time, and how does the tiny atom radiate such a long radio wave? Fortunately the observational astronomer does not have to worry about this paradoxical situation, but there is one manifestation of quantum behaviour which commands his attention, that is observed in another spectral line at the low-frequency end of the radio spectrum. This is the spectral line from the hydroxyl molecule (or more properly, radical), written as OH⁻, a combination of hydrogen and oxygen atoms, like water, H_2O, with one hydrogen atom missing. This occurs in gas clouds, such as the bright visible cloud in Orion's belt.

The line radiation from OH, at a wavelength of 18 cm, is so intense that observers could not at first explain its origin. The energy levels which determine the wavelength are related to the spin, or rotation, of the molecule and its component oxygen and hydrogen atoms.[2] An OH molecule can remain almost indefinitely in an upper energy state, until it is stimulated to emit a photon. An interesting situation occurs when this stimulation comes from another photon with the same energy, originating in another OH molecule: two photons emerge from the encounter, doubling the intensity. When there are many OH molecules in the upper energy state, waiting to be stimulated, a cascade process can occur, with more and more photons being released and adding to the intensity of the spectral line radiation. All that is needed to keep the process continuing is some way of recharging the energy levels of the OH molecules, and this happens when the OH cloud is bathed in light from a star within or close to the cloud. The process is called 'microwave amplification by the stimulated emission of radiation'. The acronym 'maser' points out the close resemblance to the more familiar laser—a similar process producing a very bright light rather than intense microwave radio.

Another molecular species which produces a very bright radio spectral line by such a maser action is water (H_2O). Maser radiation from both hydroxyl and water is so bright that it can be observed in gas clouds in some distant galaxies. These are galaxies known as 'starburst' galaxies, due to their unusually large rate of formation and explosion of stars. In the galaxy NGC 4258 the maser radiation originates in gas clouds orbiting round a central black hole, which has a mass forty million times that of the Sun. The masering gas is in a ring in the outer part of the galaxy, and rotates with it in the plane of our line of sight, so that line radiation from one side is coming towards us and from the other it is moving away. This splits the line into two parts by the Doppler effect, giving a remarkable measurement of the orbital velocity.[3] The maser emission has been mapped in detail by Very Long Baseline Interferometry (VLBI; see Chapter 9). The combination of the geometrical map and the measured velocities provides a very valuable measurement of the distance of the galaxy, contributing to measurements of the scale of the Universe.

More and More Molecules

The discoveries of spectral-line radiation from hydrogen and hydroxyl stimulated a search for lines from other molecules which might exist in gas clouds in our galaxy. In many cases the radio frequencies were already predictable from quantum theory, and most turn out to be at higher radio frequencies; that is, at shorter wavelengths. Many familiar molecules, including oxygen (O_2), carbon monoxide (CO), water (H_2O), and ammonia (NH_3) have been observed, each with many spectral lines, some of which are amplified by maser action. Water has a line-emission at 1.35-cm wavelength (frequency 22 GHz), and carbon monoxide at 2.6 mm. These can be observed in gas clouds throughout the Milky Way, and are found to be useful for determining the structure and dynamics of our galaxy. The discovery of molecules with more than two atoms was initially a surprise, since the expectation was that larger molecules would be disrupted by the all-pervasive ultraviolet radiation from stars. The reason that this does not happen is the unexpectedly large amount of dust grains in the gas clouds, which shield gas within the clouds. The most spectacular complex molecules discovered so far are carbon-chain molecules, the cyanopolyynes, such as HC_5N and $HC_{11}N$—chains of five and eleven carbon atoms with hydrogen and nitrogen atoms at the ends. Figure 58a shows cyanoacetylene, HC_3N, the first long chain to be identified, with the simpler molecules of water and ammonia, and Buckminsterfullerene, the spherical molecule C_{60} consisting of sixty carbon atoms. Even larger fullerenes have been detected, including the rugger-ball-shaped C_{70}, which rejoices in the name

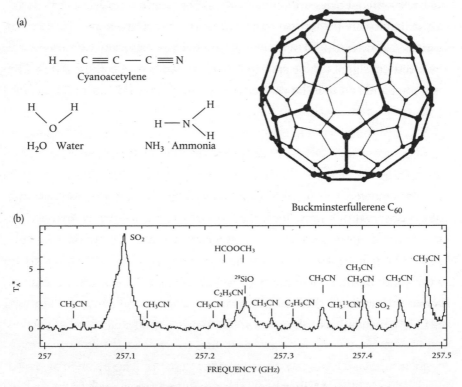

(a)

H — C ≡ C — C ≡ N
Cyanoacetylene

H_2O Water

NH_3 Ammonia

Buckminsterfullerene C_{60}

(b)

Fig. 58. (a) Some of the molecules found in interstellar space: cyanoacetylene HC_3N, water H_2O, ammonia NH_3, and Buckminsterfullerene C_{60}. (b) A spectrum of radio emission from the Orion A molecular cloud, showing molecular spectral lines.

cyanodecapentayne. Typically, these molecules exist in cold gas clouds, with temperatures ranging from around 10K to 70K. Thermal radiation at this low temperature is at millimetre wavelengths, and most of the spectral lines are therefore observed in the range 2–8 mm. The example in Figure 58b is a spectrum of emission from the Orion A molecular cloud, showing several molecular species.

Observing techniques are completely different from those for optical spectrographs, which act by splitting the light with a prism or diffraction grating. Radio waves are more easily manipulated than light waves; a radio spectrograph can act on an electrical signal to reveal its frequency structure, just as we display the frequency structure of a sound such as a bird song. The radio frequencies are often too high for a simple analysis, but there is no difficulty in transposing the spectrum down to lower frequencies, usually below 1,000 MHz (1GHz). At these comparatively low frequencies an electrical signal can be split into narrow bands by electrical filters. Each narrow band can then be converted into a stream of digits, and a detailed frequency analysis can then be carried out by a computer.

Some of the spectral lines that are so useful in millimetre astronomy are unfortunately also the source of the most serious difficulty in observing at these short wavelengths. These include emission and absorption by carbon dioxide, oxygen, ozone, and especially water vapour, since all ground-based observations must look through the terrestrial atmosphere. Astronomical radiation can only be detected in 'windows' between the worst atmospheric lines, and even at 1-mm wavelength the transmission through a typical dry atmosphere may be only 70 percent. The atmosphere is also warm and radiates thermal radiation, creating a bright sky background against which any astronomical source must be detected.

Millimetre-Wave Telescopes

As we construct radio telescopes for shorter and shorter wavelengths, we have to build reflectors with more and more accurate surfaces. A general rule is that the surface must be accurate to around one twentieth of a wavelength, so that the reflecting surface of a telescope for 1-mm wavelength must be accurate to 50 microns—about the diameter of a human hair. Furthermore, this accuracy must be maintained despite changes in ambient temperature and varying gravitational forces as the telescope moves in elevation. This is a hard specification to meet over a large reflecting surface, and individual reflectors for millimetre astronomy are for this reason mostly no more than 25 metres in diameter. The largest current millimetre-wave telescope is the 50-metre Large Millimeter Telescope in Mexico.

The requirement for surface accuracy is well known in the shorter wavelengths of the optical regime, where mirrors must be a thousand times more accurate, to some tens of nanometres. Large optical telescopes also have to deal with randomly varying refraction in the atmosphere; and millimetre-wave telescopes have a similar problem, which particularly affects large arrays. Here the signals from the individual telescopes have to be combined as interferometer pairs, and any variable delay due to the atmospheric water vapour must be compensated for as it occurs.

This difficult regime of millimetre astronomy, both using single dishes and a small array, has been explored for some years at the Institut de Radioastronomie Millimetrique (IRAM), jointly run by France, Germany, and Spain. IRAM has two observatories in southern Europe, with a 30-metre dish at Pico Veleta in Spain and an interferometer array, the Northern Extended Millimetre Array (NOEMA), presently comprising seven 15-metre (with a plan to add five more) dishes mounted on 2 km of railtrack, at the Plateau de Bure observatory in southern France. At altitudes of 2,550 metres, and 2,850 metres respectively, these observatories are usually above most of the clouds, but what is less obvious is that they are also above most of the invisible water vapour in the atmosphere, which is concentrated at lower altitudes. If the remaining water vapour above the observatory could be condensed, it would on a dry day form a layer only about 5 mm thick. Even this, however, is enough to disturb observations both in the single-dish telescopes and in the interferometer array. IRAM and other millimetre-wave observatories use receivers tuned to an atmospheric water line to monitor the 'wet' content of the atmosphere, using these measurements to estimate corrections for variable absorption and changes in propagation path length.

Single-beam observations of an extended source, such as a molecular cloud, are necessarily slow, entailing scanning of the source point by point. Following the pattern of large optical and infrared telescopes, the 30-metre Pico Veleta telescope has been equipped with an array of detectors at the focal plane, allowing multiple beams to be used simultaneously. Each detector is a bolometer—essentially a thermometer measuring energy absorbed at a wavelength of 1.25 mm. Packing multiple bolometers closely together is technically difficult, but the array for the Pico Veleta telescope has two hundred detectors, providing simultaneous multibeam operation with a beamwidth of 15 arcseconds covering a patch of sky 2 arcminutes across.

Two notable submillimetre telescopes were built on Mauna Kea, Hawaii: the James Clerk Maxwell Telescope (JCMT) in 1987, and the Caltech Sub-Millimetre Observatory (CSO) in 1985. Joining these together as an interferometer showed that an array could be built at such short wavelengths. This, with experience at IRAM and other observatories, led to proposals in Europe, the USA, and Japan for very large interferometer arrays working at millimetre wavelengths. Recognizing that these proposals all needed the best possible site, and that the cost could only be met in an international consortium, these proposals were merged into a single international proposal for ALMA, the very large array described next.

ALMA

The realization that millimetre astronomy is a powerful tool to observe molecular gas clouds in situations ranging from objects within the Solar System to very distant galaxies, opening opportunities for a new branch of chemistry, inspired the design and construction of the Atacama Large Millimetre Array (ALMA), one of the most ambitious of international scientific projects. In essence, the intention is to place the largest possible array of millimetre-wave telescope dishes on the driest possible site on Earth, and connect them as a synthesis array, producing the highest possible angular resolution (Plate 20). The dry site is in the Atacama Desert in Chile, at an altitude of 5,000 metres, where the atmospheric water vapour content is often as low as 1 mm.

The main ALMA array comprises fifty 12-metre dishes which can be arranged with a maximum extenson of 16 km, with a further twelve 7-metre antennas which can be placed closer together, and four 12-metre dishes operating singly. ALMA is designed to operate in all the atmospheric windows between 1-cm and 0.3-mm wavelengths. All sixty-six dishes need to be connected by fibre-optic links whose lengths are stable to a small fraction of a millimetre. Furthermore, the dishes must be moveable, so that the synthesized telescope can be operated either as a concentrated array for observing comparatively large objects with maximum sensitivity, or at maximum extension to provide the largest possible angular resolution. The angular resolution at wavelengths below 1 mm reaches 10 milliarcseconds, which is at least five times better than that of the Very Large Array or the Hubble Space Telescope, corresponding to a hair's breadth at a distance of 2 km.

Immense practical problems had to be overcome in constructing ALMA, not least in working at an altitude of 5,000 metres. The sixty-six dishes were assembled at a site halfway up the mountain, and were transported by road on a specially designed trailer, carrying a load of 114 tonnes (Figure 59). Observers work at the lower altitude, and only the minimum of equipment is located on site, including the computer equipment which correlates signals from each pair of dishes.

The surface accuracy of the individual dishes was demonstrated in a pilot project known as the Atacama Pathfinder Experiment (APEX), when one of the first dishes to be constructed was used over the whole millimetre range, extending to the very short wavelength of 0.2 mm. At this wavelength, even under the driest conditions in the Atacama desert, there is very large atmospheric absorption; nevertheless, the telescope was able to observe a spectral line from carbon monoxide, and discovered a line from the unusual molecule H_2D^+, a combination of hydrogen and deuterium. The APEX

Fig. 59. A 12-metre dish of the ALMA array on the transporter vehicle.
ESO/M. Marchesi

telescope has a Cassegrain secondary reflector, 75 cm in diameter, with a
train of mirrors behind the main mirror directing the beam to either of two
Nasmyth focal cabins. These contain cooled receivers, with the most sensitive parts cooled to 4K. The main reflecting surface, 12 metres in diameter,
has 264 separate aluminium panels, each with five adjustable supports.
Adjustment of the surface to produce a perfect beam shape requires accurate measurement, which is achieved by observing the beam shape and
deducing the surface irregularities from any deviation from a theoretical
perfect shape. A transmitter, wavelength 3 mm, on a distant mountaintop is
used as a source, and an auxiliary receiver is mounted forward of the
secondary mirror as a phase reference. The accuracy achieved is sufficient
for normal operation down to a wavelength of 0.3 mm.

APEX achieved a remarkable success in observations of spectral lines,
mainly at wavelengths near 1 mm. The number of different molecular types

now known in interstellar clouds is now in the region of two hundred, including many in which the atomic species are isotopes. Carbon, for example, which normally has atomic weight 12, commonly occurs with atomic weight 13. Hydrogen is observed as deuterium, which has double the atomic weight. Many of the molecules have no simple names, and following chemical convention leads to clumsy names, such as fluoro-methylidinium for the ion CF^+. Deuterated ammonia is more easily recognized; it is best written as ND_2H.

ALMA is designed to observe cold objects such as the surfaces of planets and dusty clouds in the Milky Way, as well as spectral-line emission from molecules in gas clouds and in comets. Its sensitivity is such that it has already been used to detect dust and molecular gas such as water and carbon monoxide in very distant galaxies, with redshifts up to 7.1,[4] showing that surprisingly large amounts of elements heavier than hydrogen were already present in the young Universe.

ALMA also provides very accurate positions, as demonstrated by some vital observations of Pluto. This, the most distant Solar System planet (now officially classified as a 'dwarf planet'), has a close companion, Charon, whose existence was established only in 1978 by J. W. Christy using the 1.55-metre reflector at Flagstaff, Arizona. The surface temperature of this planetary twin is 43K, which is well suited to millimetre-wave astronomy. Their apparent distance apart varies as they orbit round one another, reaching a maximum of only 10 arcseconds. A beautiful ALMA photograph shows the two objects as completely separate. The success of NASA's New Horizons spacecraft, which completed its flyby of Pluto in 2015, depended on very precise knowledge of the expected position of Pluto at this time. This was provided by ALMA measurements relating the position of Pluto to a precise reference grid of positions, which is derived from a set of very distant quasars.

Another early result demonstrates the sensitivity of ALMA both for spectral line and for wideband thermal radiation. The two stars of the binary system HK Tauri are each surrounded by gas clouds in the shape of discs. These are regarded as the birthplace of planetary systems, which might be related to the history of our own Solar System. ALMA has measured the dynamics of the two discs by observing the velocity of a carbon monoxide spectral line near 1 mm wavelength, showing an unexpected misalignment of the discs.[5] A smaller disc, which appears to show the process of planetary formation, is seen in HL Tauri (Plate 21).[6]

The Gravitational Lens

ALMA has also provided a spectacular demonstration of a natural cosmic lens which completely outclasses any of our telescope lenses, giving us a view of the most distant galaxies in the early Universe. This is the gravitational lens, the first example of which was found by Dennis Walsh (1933–2005) in a survey, at Jodrell Bank, of radio galaxies and quasars.[7] Light, or radio waves, from a distant galaxy are deflected by the gravitational field of an intervening galaxy, and can appear as a magnified and distorted image. If the distant source and the lensing galaxy are exactly aligned, the image is circular, and is then known as an Einstein ring, commemorating Einstein's prediction of the effect. Plate 22 shows an example of an almost complete Einstein ring obtained by ALMA in observations at 1 mm wavelength, with an angular resolution of 30 milliarcseconds.

Herschel and the Infrared

ALMA is at the limit of wavelengths which can be used for high-angular-resolution observations at ground level, even on the driest site at 5,000 metres altitude. Attenuation in the atmosphere is already a severe problem at 1 mm, and at ALMA's shortest wavelength of 0.3 mm only occasional observations are possible. At wavelengths of 0.3–0.2 mm there are a few atmospheric windows where perhaps 10 percent of astronomical radiation can be detected from the very driest sites (in Chile and Antarctica) on the driest few days of the year, but ground-based observing is almost or completely impossible over a decade of wavelength range reaching to the long-infrared.

This is, however, a wavelength range which matches the radiation from the coolest and dustiest clouds in which stars form, not only in our galaxy but in some of the most interesting distant and very young galaxies. The only way to see these is to observe from a spacecraft outside the atmosphere, and preferably at a large distance from the bright atmospheric radiation. Herschel (the spacecraft with a 3.5-metre telescope described in Chapter 5) covered the wavelength range 55 microns to 0.67 mm, filling the wavelength gap, but of course without the spectacular angular resolution and sensitivity of ALMA. The detectors in Herschel relate both to radio and to infrared techniques. Bolometers and photometers were used, as they would be for longer-infrared radiation, but in addition a spectrometer used the radio system of shifting frequencies down to a manageable range by mixing with a local oscillator.

Herschel operated for its full planned life, from 2009 to 2013. Its observations of cool gas extended to a previously unknown galaxy at the very high redshift $z = 6.34$, showing star formation when the Universe was only 880 million years old.[8] A successor, the Space Infrared Telescope for Cosmology and Astrophysics (SPICA), is planned, though this may not be launched until 2025. Even though the interferometer techniques which give ALMA such a high angular resolution cannot be matched in spacecraft, Herschel operated with multiple beams less than 1 arcminute across, which is well suited to the target of star-forming regions in galaxies. A quite different requirement for telescopes operating in the millimetre and centimetre range is to survey a large area of sky with smaller angular resolution but with far greater sensitivity and accuracy, in pursuit of the faint background of millimetre-wavelength radiation from the Big Bang origin of the Universe. As it happened, the most sensitive of these spacecraft designed for cosmology, Planck, was launched together with Herschel into an L2 orbit, and operated for a similar period. The objectives were, however, very different, as will be described in Chapter 11.

11

OPENING THE COSMOS

The strangest thing about the Universe is our ability to describe it, located as we are in the thin skin of an insignificant-sized planet inside a bewilderingly huge cosmos. Although its extent in time and space is so far from our own local experience, we can describe with some precision how the Universe developed from a dense fireball, expanded, and continues to expand on a measurable time-scale. In contrast, we know much less about the smaller details: how exactly the stars and galaxies formed, and how we, the observers, are related to the complexities of the more local Universe. In this chapter we will describe how we can see beyond the complexities and describe the cosmos as a whole with remarkable precision and simplicity.

The Distant Nebulae

A century ago the known Universe stretched only as far as our Milky Way galaxy. Nothing further was known to exist. The nature of our galaxy as merely one of an immense number of similar galaxies was a matter for speculation only. The break-through came in 1912 when Vesto M. Slipher (1875–1969) used the 24-inch Clark refractor at Flagstaff, Arizona, to measure the spectrum of light from several nebulae—faint patches of light which were thought at that time to be gas clouds in our galaxy. He found instead the characteristic spectra of stars, very faint and therefore possibly very distant. Furthermore, he found that the spectra were redshifted, showing that these nebulae were moving away from us at very large recession velocities.[1] Of twenty-four nebulae, twenty-one were moving away from us with velocities up to 1,000 km per second, far greater than any velocities known within the Milky Way. In 1929, Slipher (always known as 'V. M.') published redshifts for 465 others, most of which were moving away (though the Andromeda Nebula turned out to be an exception, moving towards us at 300 km per second).

The idea that these distant nebulae were part of a uniformly expanding Universe was first set out by Georges Lemaître (1894–1966), Belgian cosmologist and Catholic priest. Using apparent brightness as a measure of distance, he introduced the idea of a uniformly expanding Universe, in which the whole framework of space in which the distant galaxies were embedded was expanding. The more distant was any galaxy, the faster would it be moving away from us. His 1927 paper, which is a rich combination of theory and observation, is recognized as the first exposition of Hubble's Law.

Edwin Hubble (1889–1953) arrived at Mount Wilson observatory in 1919, coincidentally when the 100-inch Hooker Telescope was completed. This was the ideal instrument for reducing the uncertainties in the distances of the extragalactic nebulae, which he famously achieved by resolving individual stars in the nearest and brightest. A comparison with the brightness of similar stars in the Milky Way gave distances which he used in a 1929 paper, giving an improved linear relation between distance and recession velocity. The ratio of velocity-to-distance became known as the Hubble Constant.[2]

Hubble is usually celebrated as the discoverer of the expanding Universe. Lemaître's earlier 1927 paper,[3] published in French in Brussels, was not widely read, and its significance was only widely appreciated several years later.

Lemaître pointed out that the observed expansion of the Universe could be tracked back in time, implying that everything originated in a single primordial concentration at a definite time in the past. As the interval between this past time and now is inversely proportional to the Hubble Constant, it became generally known as the Hubble Time. (This is not the modern estimate of the age of the Universe, as observations now extend very much further and the geometry of the expansion is better understood, giving a remarkably precise value of 13.8 billion years.) Hubble went on to

use the 100-inch telescope to distinguish the two main classes of extragalactic nebulae—spirals and ellipticals—and made the first estimates of their
population in space. He used a simple test to show that those within his
reach were distributed evenly with distance, by counting how many were
seen at each level of brightness. This test was repeated several decades later
for radio galaxies, when it could be extended to much greater distances,
where the effects of evolution in the Universe were becoming apparent.

Further Back in Time

The galaxies observed by Slipher and Hubble were, by modern standards,
not far away; the redshifts were no more than 0.1. Much larger redshifts
were only discovered almost half a century later. When quasars were first
identified, redshifts greater than 1 were found, and modern cosmology was
born. These large redshifts implied that the whole Universe was expanding,
and had more than doubled in size since the light was emitted by these
quasars. We were looking back to a time when the Universe was only half
its present age. The record highest redshift has grown to over 8, for the
most distant identified source of gamma-ray bursts. At such large distances,
and looking back so far towards the beginning of the expansion, the
geometry of space and time needs careful definition. However, geometrical
theory now has to be expanded to take account of an observation of a very
much earlier stage of the developing Universe, at redshift $z = 1,000$. This
discovery of the Cosmic Microwave Background (CMB), which we now
describe, transformed cosmology.

The Discovery of the Cosmic Microwave Background

In 1933, when Jansky discovered radio emission from the Milky Way, his
research was part of investigations of long-distance communications by the
Bell Telephone Laboratories (BTL). Thirty years later, BTL was again

exploring the basic limits of radio communication, but at much shorter wavelengths and involving relays or echoes from satellites. As before, the aim was to measure the background of radio noise against which radio communications must contend, but at wavelengths around several centimetres rather than tens of metres. At centimetric wavelengths there is very little radio emission from the Milky Way. Using temperature as a measure of the radio brightness of the sky, a receiver was needed which could measure a temperature of only a few degrees absolute. Only a simple telescope was needed, but it was essential to avoid radiation from all warm objects, especially the surrounding ground. For this measurement, BTL built the telescope shown in Figure 60, and set two radio astronomers, Arno Penzias (b.1933) and Robert Wilson (b.1936) to measure the temperature of the sky. The telescope was a 20-foot horn, designed to have a very clean beam which

Fig. 60. The big horn telescope at Bell Telephone Laboratories, with which Arno Penzias and Robert Wilson discovered the Cosmic Microwave Background.

would avoid all unwanted radiation from the surroundings. In 1964, Penzias and Wilson found background radiation from everywhere in the sky, at a temperature of 3K. First, they first had to eliminate any possibility that this might originate in their receiver, or in the horn itself, where pigeons had roosted and fouled the surface. Cleaning everything made no difference; the radiation they had indeed discovered was truly cosmic.

The origin of this 3-degree microwave background was soon recognized by Robert Dicke (1916–1997), who was independently working at Princeton to detect the cooled remnant of the hot young Universe. Bizarrely, Penzias and Wilson did not know that this remnant radiation, now known as the Cosmic Microwave Background, had been predicted, while Dicke knew what to expect but had not yet found it. The two research groups were introduced to one another by another radio astronomer, Bernard Burke (b.1928), who can be regarded as a midwife at the birth of a new phase in observational cosmology.

In the thirty-year gap between Hubble's discovery of the expanding Universe, which suggested an origin in some sort of primordial concentration, and the discovery of the CMB, there was one theoretical investigation which produced a clear picture of how the young Universe might be observable. In 1948, Ralph Alpher (1921–2007) and Robert Herman (1914–1997) took the concept of the expanding Universe to its logical conclusion, or rather its beginning, by working out what must have happened when the whole Universe was starting its expansion.[4] The young Universe would be very dense and very hot, and would be expanding rapidly, and cooling as it expanded. As it cooled, two things happened in the young Universe which might have observable consequences at the present day. First, some light elements, particularly helium, should be created from the original elementary particles; the proportion of helium in the Universe today does indeed match this prediction. Second, the hot ionized gas should cool as it expands, and should become cool enough to start forming neutral molecules. This stage is known as the Epoch of Recombination.

As our telescopes look further and further into the distant Universe, we are also looking further and further back in time. There is no possibility of looking further back than the Epoch of Recombination, since no light nor any other electromagnetic radiation can penetrate the dense ionized gas of the young Universe before that era. However, the neutral, recombined gas, mainly hydrogen and helium, of the adolescent Universe is transparent, so we can look through it and measure the properties of the Universe at the Era of Recombination. What we see is the Cosmic Microwave Background.

At that stage of expansion the Universe had cooled to a temperature of around 20,000K—not much hotter than the atmospheres of many stars in the Milky Way. At this distance the Universe was still young, and space itself, along with everything in it, was far more condensed than our local Universe. Light reaching us from such a distance is transformed along with the expansion of space, and reaches us with a wavelength stretched by a factor of 1,000. This is a very large redshift, far higher than the redshifts of local galaxies which gave the first glimpse of the expansion. It has a fundamental effect on the intensity of the radiation, as well as on its wavelength. We measure the intensity as a temperature: as the wavelength stretches, so does the temperature fall.

If the wavelength of light reaching us today has been stretched by a factor of 1,000, then the temperature of the background radiation should be only a few degrees above absolute zero. (A good way of comprehending this is to imagine a photon, whose energy falls as the wavelength increases, travelling all the way across the expanding Universe from the hot gas to our telescope.) This was the prediction which was so dramatically confirmed by Penzias and Wilson in 1964. The CMB became known as the Three Degree Background.

The first detection of the CMB, by Penzias and Wilson, was at 7 cm wavelength, but the peak of thermal radiation at a temperature of 3K should be closer to 1 cm, and it should even be detectable at shorter wavelengths, extending to the long-infrared. Measuring the shape of the spectrum would provide a conclusive check on the theory, but sufficiently precise infrared observations can be achieved only above the atmosphere, avoiding the absorption and radiation from atmospheric water vapour. This was achieved by a spacecraft, the Cosmic Background Explorer

(COBE), which was placed in near-Earth orbit in 1989, carrying a Far Infrared Astronomical Spectrograph (FIRAS). This spectrograph was spectacularly successful, measuring the spectrum of the CMB over the wide wavelength range from 1 cm down to a 0.1 mm.

The FIRAS spectrograph followed a principle devised by A. A. Michelson (1852–1931)—the same pioneer who invented the interferometer described in Chapter 9. In the FIRAS spectrometer, infrared radiation is collected by a small horn, covering 7° of sky. Between the horn and the detector there is a pair of mirrors with a variable spacing between them; this is the interferometer. A motor drives one of the mirrors, producing a spacing up to a maximum of 6 cm, and the spectrum can be found by the way in which the light varies at the detector.[5] The whole of this apparatus has to be kept very cold, at 1.5K, to avoid thermal radiation. The 7-degree beam was scanned across the sky by the slow rotation of the spacecraft.

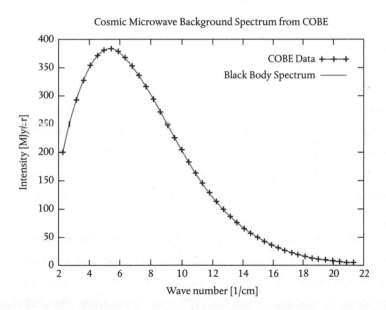

Fig. 61. The spectrum of the Cosmic Microwave Background, as measured by the Far Infrared Astronomical Spectrograph (FIRAS)—the spectrometer on the Cosmic Background Explorer (COBE). The observed points fit exactly on the smooth curve, which is an ideal spectrum of thermal radiation at a temperature of 2.726 K.

The astonishingly accurate spectrum of the CMB measured by COBE is shown in Figure 61. The measured points fall exactly on the solid line, which is the expected theoretical spectrum for thermal radiation at 2.726K. The spectrum, and the intensity, is the same over the whole sky, except for a large-scale so-called dipole component, seen as a slightly increased temperature in one half of the sky and a reduced temperature in the other half. This component is attributed to the Earth's motion in relation to the whole cosmos, at a velocity of 369 km per second; the sky appears slightly warmer in the direction of motion, and cooler in the opposite direction, by 3 millikelvins (0.003K)—that is, about 0.001 of the average brightness. Apart from this dipole component, the sky seemed uniformly bright.

The Early Universe: Ripples in the Cosmos

How could the featureless CMB evolve into the complex Universe that we know today? Our Universe is obviously very far from smooth, containing galaxies and clusters of galaxies which must have grown from structure in the primordial Universe. There must be some ripples in the smooth surface of the background, which grow as everything expands. In fact, some of this structure was found by another instrument on COBE, the Differential Microwave Radiometer (DMR), which was set up to look for structure at an angular scale of a few degrees. There have since been many more accurate and detailed observations, from spacecraft, from balloons, and from ground level, which have dominated observational cosmology ever since and given us a remarkable detailed understanding of the way in which our Universe came into being.

The first detailed picture of the structure of the CMB was obtained by BOOMERANG, the unforgettable name of a balloon-borne radiometer which flew for ten days at an altitude of 39 km over Antarctica. Like the DMR, the radiometer in BOOMERANG compared the sky brightness in two opposite directions, scanning the sky as it slowly rotated. The ten days allowed a long and precise averaging, or integration, and was also very convenient for the balloon launchers, as the balloon was in an air current which circulates round the South Pole, bringing it back close to the launch site for its descent (Plate 23).

BOOMERANG provided the first clear map of the structure of the CMB,[6] showing predominant ripples on a scale of 1°. This was exactly the angular scale which had been predicted by cosmology theory; furthermore, theory predicted that there should also be structure on a series of scales close to 0.5°, 0.25°, and so on, forming an harmonic sequence. The exact angular scales, and the amplitudes of each harmonic, would reveal the details of dynamic processes in the early Universe, before and during the Era of Recombination. More observations would be needed for this, using telescopes with narrower beams. The immediate need, however, was to improve the accuracy of the maps of the larger-scale ripples. This was achieved by observations from two dedicated spacecraft: WMAP and Planck.

The ripples in the CMB are small, at a level of less than 0.001 percent of the 3K background. Measurements at this level require very stable detectors with very low intrinsic noise, and long integration times. Spacecraft provide ideal platforms, since they can operate at low temperatures and undisturbed for many days or even for years. Their main advantage is in their freedom from radiation and absorption in the Earth's atmosphere, and they can also avoid other interfering sources of thermal radiation, particularly the Sun and the Earth. The Wilkinson Microwave Anisotropy Probe (WMAP)[7] observed from a distance of 1.5 million km from Earth, orbiting slowly about the Lagrangian L2 point (see Chapter 5). It scanned the sky with a pair of 1.6 x 1.4-metre identical telescopes pointing in opposite directions, measuring the small differences as the spacecraft rotated every two minutes. COBE had operated at long-infrared wavelengths, while the WMAP telescopes operated at radio wavelengths from 13 mm down to 3 mm (23–94 GHz). At these wavelengths, in contrast to infrared, sophisticated electronics can be used to measure small differences in the detected power.[8] Furthermore, it turned out that making measurements at several different millimetre wavebands is essential for distinguishing between the CMB and some overlying weak radiation from the Milky Way. This background radiation from our galaxy includes contributions from warm dust, ionized hydrogen, and synchrotron radiation from high-energy electrons, all of which have different spectra. They can be distinguished and removed by making maps at different wavelengths.

WMAP was an outstanding success, and its results have been acclaimed as the beginning of precision cosmology. Calibration of receivers working to unprecedented low-signal levels, measured in microkelvins, used a new standard, the dipole component due to the Earth's motion, whose temperature of 3 millikelvins was measured by COBE. The geometry of the telescope beams was checked by observations of thermal radiation from the planet Jupiter. WMAP provided precise measurements of the scale and amplitude of the ripples in the CMB, on angular scales down to a fraction of a degree. The remaining problem was to fill in the fine detail of the ripples, at the smaller angular scales.

The Fine Structure

The difficulty in measuring the small-scale ripples was that telescopes with narrower beams must have larger apertures, measured in hundreds of wavelengths—larger than can be launched on a spacecraft. The peak of the CMB spectrum is at a wavelength of 2 mm, so a telescope with an aperture of at least several metres is needed for the finest structure. This challenge could be met only by ground-based telescopes. Despite the difficulties of observing from the ground, a whole series of specialized telescopes have been built for operating on the best dry sites, aimed at measuring the size of the ripples on a scale reaching down to a few arcminutes. The preferred technique for the first of these was familiar in radio astronomy: looking at fine structure needs an interferometer. The overall size to produce sufficient angular resolution at such short wavelengths need be no more than 10 metres, which can be achieved by mounting some tens of small dishes close together on a simple steerable mount. The first, built on Tenerife in the Canary Islands, was actually named the Very Small Array. Another, the Cosmic Background Imager (CBI), was built on the high Atacama desert site in Chile. The CBI used thirteen dishes, each 1.4 metres in diameter, on a single mount (Plate 24).

The interferometeric telescopes successfully extended the measurements down to a scale of 10 arcminutes. Surprisingly, the next generation of ground-based telescopes designed to measure the structure of the CMB reverted from the interferometer technique to a measurement of total power in a narrow beam. An example is the South Pole Telescope (SPT)—a radio telescope built in 2007 and initially working at 2 mm wavelength. A single beam is almost useless, since the low signal levels require long observation times to beat down the noise level. The SPT uses an array of 1,000 detectors in the focal plane, producing 1,000 beams, each 1 arcminute in diameter and together covering 1 degree square of sky. The telescope is a Gregorian with a 10-metre dish made of 217 panels, with a second reflector off-axis as in the original optical Gregorian (see Chapter 1). As in all such Gregorian configurations, the field of view is increased by including a weak lens. For microwave radio a polythene lens is used, with the added advantage that polythene acts as a barrier against unwanted infrared radiation.

Each detector in the array is a tiny horn only 150 mm in diameter, feeding a superconducting transition edge sensor (see Chapter 4). The whole tightly-packed array has to be kept at a temperature of only 0.5K, which fortunately is now possible with a continuously acting helium refrigerator. The telescope is isolated from all local sources of thermal radiation by a reflecting screen, so that the sky itself is the only source of thermal noise to reach the detector array. The sky noise is dominated by atmospheric radiation, but at the South Pole site at an altitude of 2,800 metres the water vapour content is the lowest in the world; it has the low median value of only 0.25 mm (the depth if it were all precipitated on to the surface).

Constructing the SPT was a formidable logistic exercise. The total weight of the telescope is 244 tons, and every part of it had to be flown to the

South Pole research station on a transport aircraft capable of carrying only 11-ton loads. The panels of the dish had to be adjusted *in situ* to an overall accuracy of 20 microns, which required accurate measurements of beam shape (including phase) from a locally mounted transmitter. Stability in performance is crucial, and the telescope is designed to have no moving parts inside. Continuous observations are made during the six months of winter, when atmospheric conditions are ideal.

The South Pole Telescope has broader objectives than delineating the structure of the CMB; it is also intended to search for clustering of galaxies, which is, as we will see later, a closely related subject in cosmology. The main contribution so far has been the extension of the structure of the ripples down to a scale approaching 1 arcminute, averaged over most of the southern sky. The results have been combined with those of the most recent spacecraft, Planck.

Planck: the Supreme Cosmology Spacecraft

Although COBE and WMAP are rightly celebrated as the spacecraft which initiated precision cosmology, there remained some questions to be answered. As WMAP had shown, some of the ripples measured in the microwave sky originated in the foreground of radiation from the Milky Way. These could be distinguished only by measurements covering a wider range of wavelengths, from infrared through to longer millimetre waves. And greater sensitivity was already known to be possible, as demonstrated by balloon-borne telescopes using new bolometric techniques. These culminated in Archeops—a series of balloon flights carrying detectors cooled to the astonishingly low temperature of 0.1K. In these detectors, radiation is focused onto a spider web of fibres only 1 micron thick, coated with gold. At the centre of the web is a tiny thermistor, whose electrical resistance varies with temperature (a measuring device working on the same principle as familiar domestic thermometers). The last Archeops test, which took place in 2002, flew from Kiruna, Sweden, to an altitude of 32 km. Its scan of a small portion of sky showed a much greater sensitivity than WMAP, and actually preceded the WMAP results in a delineation of the cosmic ripples over 30 percent of the sky. The European Space Agency decided to use the new technique in a major spacecraft, named after the famous physicist Planck, which was launched in May 2009 on the same launch vehicle as Herschel, the infrared spacecraft.

The gap between infrared and radio which the atmosphere imposes on ground-based astronomy no longer exists for spacecraft; the wavelengths covered by the infrared spacecraft Herschel (Chapter 5) extend to a wavelength of 0.5 mm (600 GHz), while Planck—basically a microwave radio observatory using longer wavelengths—extended to the short wavelength of 0.35 mm (857 GHz). Even the techniques overlap; although at the lowest frequencies of 30, 44, and 70 GHz Planck used normal radio techniques with transistor amplifiers, in its higher frequency bands, 100–857 GHz, the receivers used a technique much closer to infrared than radio; this was the spider-web thermistor tested in Archeops. Between them, Planck's nine bands stretched over a 30-to-1 range of wavelengths, from 1 cm downwards. All the detectors were fed by horns arranged on the focal plane of a single mirror with a 1.5-metre aperture. The spacecraft rotated about an axis directed away from the Earth and the Sun, taking two minutes to scan a narrow strip of sky. The whole sky therefore took a year to cover, and only two complete scans could be accomplished before the liquid helium coolant was depleted.

The wide wavelength range of Planck was essential to sort out the several different components of the millimetre sky. At the longest wavelengths there is a significant contribution from synchrotron radiation from the Milky Way; the small residual at shorter wavelengths can be deduced from its known spectrum. Similarly at shorter, near-infrared, wavelengths there is a contribution from warm interstellar dust, which has a different spectrum. Analysing the full sky maps and removing these contributions was a major task, the success of which may be judged from the sky maps shown in Plate 25. These maps cover the whole sky, arranged so that the plane of the Milky Way runs horizontally along the centre. The uncorrected map shows the full sky brightness, including synchrotron and dust radiation from the Milky Way. Constructing the map of the pure CMB over the whole sky required the removal of these contributions in a painstaking analysis.

The structure of the ripples could now be characterized on angular scales down to 5 arcminutes, extending down to 1 arcminute by adding the measurements from the South Pole Telescope and the Cosmic Background Imager. The result is presented as a graph of ripple amplitude against ripple scale, as in Figure 62. This graph of the size of ripples in the CMB is an astounding achievement. Not only does it represent the main results of some of the most sophisticated and complex experiments ever made; it is an amazing verification of predictions made by cosmologists describing the development of the cosmos following its origin in a single catastrophic event, the Big Bang. By comparing the amplitudes and angular scales of the peaks with theory, a precise description of the Universe has been produced. The fit between theory and observation is shown in Figure 62, where the experimentally determined points are superposed on the smooth line of the theoretical model.

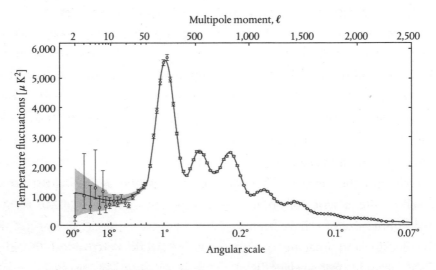

Fig. 62. The size of the ripples in the Cosmic Microwave Background vary at angular scales from several degrees down to several arcminutes.

Polarization

A further challenge now faced the observers. Having discovered the CMB, at a temperature of only 3K, followed by a delineation of the ripples at a level smaller again by a factor of 100,000 or more, they now had to stretch the sensitivity by a further factor of 1,000 to measure a small extra characteristic of the ripples: their polarization.

The polarization of light is familiar to fishermen and motorists, who use polaroid glasses when looking at light reflected from a water surface or a wet road. In all electromagnetic radiation, the electric field can be in any direction transverse to the line of propagation; the radiation is polarized when the field is predominantly in one direction. Radio and television antennas have to be oriented either horizontally or vertically to pick up man-made signals, which are usually strongly polarized, but the polarization of the CMB is very weak and must be detected by a careful comparison of the signals in two receivers picking up two components of the electric field of the radiation. In Planck, each tiny horn of the radio detectors contained two probes connected to separate receivers. The infrared spider-web detectors were made sensitive to polarization by a gold coating only the web fibres in one direction, so picking up only one polarization, and a second spider-web, with the gold-plated fibres oriented at right angles, picked up the other. The tiny difference between the two receivers was measurable only after the long averaging time of a spacecraft observation. However, the importance of polarization measurements is so great that some ground-based observations are also being made that challenge and even exceed the sensitivity of Planck.

Following the success of the South Pole Telescope, and using the same excellent site, a series of observations of polarization of the CMB at the main ripple size of 1° have been made by a 26-cm telescope. The project is the Background Imaging of Cosmic Extragalactic Polarization (BICEP). Concentrating on wavelengths at the peak of the CMB spectrum and a restricted region of the sky, and using a multiple array of detectors, BICEP reaches an extreme of polarized sensitivity. The detector array uses the transition edge superconductor (TES, see Chapter 4) at a temperature of 0.25K. BICEP2 used 512 such detectors, and BICEP3, which started observing in 2015, uses 2,560 detectors—1,280 for each polarization. A similar system, POLARBEAR,[9] located at high altitude (5,200 metres) in the Atacama desert, has been operating since 2012.

Polarization in the CMB occurs by scattering—a phenomenon well known in the terrestrial sky where sunlight scattered in the atmosphere is so polarized that bees can navigate by using the direction of polarization. In the CMB it is the background itself which is scattered in the ripples, producing a predictable pattern of polarization. There is also a more subtle pattern, known as the B mode, in which the direction of polarization follows another pattern, predicted to arise from gravitational waves generated in the early Universe. This is particularly interesting in cosmology, as the existence of such gravitational waves is closely related to a stage in the very early development of the Universe known as inflation, though it is particularly difficult to demonstrate. The difficulty lies in the foreground of Milky Way radiation which overlies the CMB. There is a magnetic field pervading the Milky Way, which polarizes the radiation, and Planck has in fact been able to delineate the magnetic field of the galaxy by measuring the polarized radiation. The measurement and removal of this foreground radiation is crucial to the detection of the vitally important B mode. An initial claim in 2014 that the B mode had been found turned out to be the result of insufficient correction in the observations. This research field is open to many more observations, mainly by ground-based telescopes and balloons, since further spacecraft improving on Planck are not proposed at present.

The Theory of the Cosmos, and Two Puzzles

There is a so-called Standard Theory in cosmology, which brings together the large-scale features of the expanding Universe and the material it contains. Newton's theory of gravitation provides a useful start: the gravitational pull of the matter in the Universe tends to shrink the material together, slowing down the expansion which started at the Big Bang. If there is enough material, the expansion will slow down, stop, and go into reverse; otherwise it will continue for ever. But we need Einstein to set out the geometry correctly. In his General Theory of Relativity he showed that it is better to represent the gravitational effect of matter as a distortion of space, which we can observe in its effect on light and radio as well as on matter, as demonstrated by the gravitational lens (Chapter 10).

The Standard Theory gives the age of the Universe as 13.8 billion years. Its age when it became visible as the CMB was only around 370,000 years, and after that time it should be possible to observe in some way the development of the CMB structure into stars, galaxies, and clusters of galaxies. The galaxies which we can see should show how the universal expansion is proceeding. Here we meet a serious snag. Following the redshifts of galaxies out to redshifts up to $z = 2$, very much further than Hubble could, shows that the expansion is not slowing down as it should if it is acted on only by gravity. There seems to be an accelerating force which becomes more important as gravity becomes weaker in the expanding Universe. The Standard Theory accounts for this by adding a new component: Dark Energy—an all-pervading source of pressure which tends to counteract gravity. The origin of Dark Energy is not understood, but its importance is emphasized by expressing its density in terms of matter. Energy and matter are interchangeable; the embarrassing fact is that the matter equivalent of Dark Energy is much larger than the total other known contents of the Universe.

The discovery of the acceleration required observation of the light from a type of supernova (a Ia supernova), which was bright enough to be seen at a large redshift. Catching the supernova in time to measure its brightness and redshift was an exciting project which used survey telescopes such as Swift (Chapter 6) and the Sloan Digital Sky Survey (Chapter 3), followed by observations with the largest optical telescopes which were available. The result led to a Nobel Prize shared by both groups.[10] The existence of Dark

Energy is made more certain by the WMAP and Planck measurements of the ripples in the CMB.

Adding to the embarrassing necessity of Dark Energy, there is another apparently essential component of the Universe as we understand it today. Most galaxies are rotating, and like our Solar System, the speed of rotation in any part should be related to the gravitational pull of everything within the orbit. But the outer parts of galaxies are all moving faster than we would expect from the observable matter. There is an extra component, totally invisible, which must be added. Its nature is unknown; it simply has a name: Dark Matter.

We now arrive at a bewildering picture of the Universe. The beginning in a Big Bang, the expansion and cooling to produce the observable CMB, the ripples in the CMB which are the seeds from which the galaxies grow—all this fits beautifully into a remarkably simple and accurate picture. But we are left with the puzzle: ordinary matter constitutes only 5 percent of the total contents of the Universe. Another 27 percent is Dark Matter, and the remaining 68 percent is Dark Energy. We have an excellent description of one twentieth of the Universe, and none at all of the remainder.

There is a large gap in our knowledge of cosmological evolution between the Era of Recombination, when the CMB was generated, and the most distant galaxies seen by our most powerful telescopes. In terms of redshift, this is a gap between $z = 1,000$ and $z = 10$. Very little is known concerning the time during which the almost uniformly dense primordial Universe formed into clusters of galaxies. The process is affected by Dark Energy, so the structure of these clusters should vary with distance as the relative effects of expansion, gravitation and Dark Energy develop. The galaxy surveys described in Chapter 3 do show such evolution: a clumping of galaxies at a scale of around 100 megaparsecs has been discovered, which can be related to the dominant scale of ripples in the CMB. What is lacking is any observation of the way in which the structure of the Universe is evolving, over a large range of redshifts from $z = 1,000$ downwards. There are, however, already several attempts, using radio telescopes to look at the behaviour of hydrogen at moderate redshifts.

The Young Universe Develops

The Universe as we know it today is very different from the almost perfectly smooth CMB. The tiny ripples in the CMB are the first manifestation of the structures which develop into clusters of galaxies, which in turn produce stars, planets, and ourselves. The way in which the ripples condense into galaxies is far from clear. The ripples originate in the very early Universe, which is opaque and invisible. The ripples become visible only at the Era of Recombination, when the Universe is cool enough for ions and electrons to combine to form neutral hydrogen. At this stage onwards the Universe consists mainly of hydrogen, and the transformation of ripples into clusters of galaxies might be traced by observing the structure of the condensing hydrogen in these early years. The only way this can be done is to observe the hydrogen radio spectral line. This is the 21-cm spectral line which has been so useful in delineating the shape of the Milky Way galaxy, but for the developing cosmos it would have to be observed with a very large redshift, starting at $z = 1,000$ and diminishing as the age increases from 370,000 years to a few billion years, when direct optical observations can take over. The largest redshifts take the hydrogen line to frequencies far below the available radio spectrum, but a redshifted hydrogen line between $z = 10$ and $z = 1$ should be observable. At $z = 10$ the wavelength would be stretched from 21 cm to around 2 metres, at frequencies around 150 MHz. As might be expected, this possibility is being taken as a challenge by radio astronomers. The search for cosmological hydrogen is being pursued in some very ambitious projects, requiring new radio telescopes devoted to metre wavelengths.

The expected metre-wavelength signals from cosmological hydrogen are very weak, and must be distinguished from the more powerful synchrotron emission from the Milky Way. Distinguishing the early hydrogen from this Milky Way background can be achieved only by looking in detail at the spectrum over a wide range, since this should show a changing pattern of structure: different frequencies arise from hydrogen at different redshifts. The Low Frequency Array (LOFAR) and the precursors of the Square

Kilometre Array (SKA, see Plate 26) offer a real possibility of achieving this measurement. It seems likely, however, that a smaller telescope dedicated solely to this task might have an earlier chance of success. An example is the Canadian Hydrogen Intensity Mapping Experiment (CHIME), shown in Figure 63.

CHIME is aimed at mapping hydrogen in the evolving Universe between redshifts $z = 0.8$ and $z = 2.5$, at observing frequencies between 400 and 800 MHz. Overall it has a square aperture of 100 x 80 metres, constructed of four cylindrical reflectors, but each cylinder has a row of separate receivers along the focal line. There are 2,560 of these individual receivers, but instead of their all being connected to produce a single beam they are

Fig. 63. The Canadian Hydrogen Intensity Mapping Experiment (CHIME) radio telescope at Penticton, Canada. The four cylindrical reflectors are aligned north–south. Along the focus line of each of the reflectors are 512 individual receivers. There are no moving parts, and each receiver can be correlated in pairs with any other.

connected in pairs as elements in a synthesis array. By connecting the receivers in pairs, using a massive multiple correlator system, an array of multiple beams is created, all beams observing simultaneously. This system rejects the total power which would be received by a single beam, and detects only the angular structure in the sky. There are no moving parts in the whole telescope, and the observations can be repeated in an unbroken session for a year or more, giving the necessary averaging time for an extremely high sensitivity.

CHIME will cover a large part of the northern sky. There are already proposals for other telescopes dedicated to observing the hydrogen in the developing Universe. The southern hemisphere is an obvious target, and there are many possibilities for filling in the gaps in the redshifts by extending the range of redshifts to $z = 10$ or more. The Square Kilometre Array (SKA) may be useful for this, but a dedicated close-packed array may be more appropriate. An example is the Hydrogen Epoch of Recombination Array (HERA), to be built in South Africa in collaboration with American universities, following experience with smaller arrays already operating in South Africa and Australia.[11]

Where Next with Cosmology?

Telescopes like CHIME and HERA are devoted to very specific cosmological questions, as indeed were COBE, WMAP, and Planck. This is a far cry from the conventional telescopes discussed in the earlier chapters of this book, which were built to explore the sky without any closely specified target. In Chapter 12 we will see that the more general approach of ever larger and more complex telescopes will complement the smaller and more specialized telescopes in our attempts to explore the many mysteries of the cosmos.

12

THEN, NOW, AND TOMORROW

Science is always looking for ways to describe the natural world in simplified terms, bringing together apparently separate phenomena in a process of unification. Biology, for example, was transformed by Darwin's theory of evolution, which brought together the whole living world; and in physics it was Faraday's observations and Maxwell's theory that unified electromagnetism.

In astronomy, the first such simplification was Galileo's demonstration that the planets are in orbit round the Sun, not the Earth, and that the Milky Way is made of stars like the Sun. Galileo achieved this by building a new telescope, without any preconceived ideas about what he would find. Serendipity, aided by an adventurous spirit and determination, rewarded him amply. The discovery of Jupiter's moons marked the start of dynamical astronomy, and the beginning of Newtonian mechanics.

The theory of the expanding cosmos, formulated by Lemaître and by Hubble, brought together the observations of the recession of the nebulae and Einstein's General Relativity, and laid the foundations for the Standard Model of the Universe which so satisfactorily describes the cosmos on the largest scales of time and space (Chapter 11). Again, the key observation to start the integration was the serendipitous discovery of the Cosmic Microwave Background. The Standard Cosmological Model was overwhelmingly adopted, and the present amazingly successful modern era of cosmology began.

Observational astronomy today is in the midst of an unprecedented boom, both in existing telescopes and in planned new telescopes. The preceding chapters trace the technical advances which have enabled a transformation in all branches of astronomy, from the Solar System through our galaxy, other galaxies at greater and greater distances, and clusters of galaxies, through to the Era of Recombination. The main factors in this transformation are the development of detectors which operate almost, or in some cases at, the level of individual photons, and the era of fast electronics and massive computation. The scale and organization of modern astronomy have also changed dramatically, matching the new opportunities for research and also relying increasingly on international cooperation and associations.

The Major Telescopes Today

Leaving aside the many specialized telescopes designed to address specific cosmological questions, as described in Chapter 11, there are in operation major telescopes operating in all parts of the electromagnetic spectrum. Furthermore, we should include in the list several based on spacecraft which are no longer operational, since they have provided invaluable catalogues and databases which are openly available and which continue to be actively mined for information.

We start at the highest energies, with the gamma-ray telescopes (Chapter 7) which comprise both the space-based Fermi LAT and the ground-based shower detectors High Energy Stereoscopic System (HESS), Very Energetic Radiation Imaging Telescope System (VERITAS), and Major Atmospheric Gamma-Ray Imaging Cherenkov (MAGIC). Practically every photon received by these telescopes is recorded, with measured values of energy and direction of arrival. At lower gamma-ray energies, Fermi LAT is sensitive over a huge solid angle of sky, and scans over the whole sky every day. It is already possible to search through the records and discover pulsars not previously known from radio surveys. In 2014, a list of pulsars detected by LAT had reached 161, including forty-one which were actually discovered by searching through the LAT database. This is a very remarkable achievement, considering the long intervals between the arrival of individual gamma-ray photons. With the first pulsar discovery, photons were detected only at the average rate of one photon per minute, even though the basic pulse period for this pulsar is less than a third of a second.

The Fermi LAT third catalogue of gamma-ray sources, published in 2015, contains more than 3,000 sources, many of which have been identified as active galactic nuclei, quasars, pulsars, and supernova remnants—all objects which are radiating with far more energy than is possible from a simple thermal source. The gamma-ray telescopes are proving that gamma rays with energies of 25 GeV and upwards originate from many of these non-thermal sources. The Gamma Ray Burst Monitor (GBM), on the same spacecraft as Fermi LAT, is also providing a large record of gamma rays from novae and other, as yet unidentified, sources.

At the next lower energy range of X-ray astronomy, the two major spacecraft telescopes XMM-Newton and Chandra have been in operation since 2000, and are likely to continue for several more years, possibly until after 2020. The 0.5-arcsecond accuracy of Chandra and the sensitivity of XMM-Newton provide well for observations in the energy range up to 10 keV, while the Nuclear Spectroscopic Telescope Array (NuSTAR) extends the range up to 70 keV. These represent the highest energies that can be imaged using a reflector technique, while individual transient sources at higher energies can be located using a coded mask technique, as with Swift.

The lowest X-ray energies merge into the extreme ultraviolet (EUV), where the Hubble Space Telescope takes over, providing spectroscopy with the Cosmic Origins Spectrograph (COS), installed on a servicing mission in 2009. The wavelength coverage of COS is 320–115 nm, extending from the visible as far into the shortest wavelength ultraviolet light as can be achieved in a telescope which is basically designed for visible light. COS observes the spectra of faint point-like objects in the Milky Way and active galactic nuclei.

The visible spectrum, and its extension into the near-infrared has, from the time of Galileo, always been the most productive spectral range, and remains so for several reasons. Most obviously, the atmosphere is transparent (in good weather), so that large optical telescopes can be built on dry high-altitude sites such as the Atacama desert in Chile. Detector technology has also developed to provide us with arrays with very large numbers of pixels, working with such efficiency that most of the photons arriving at a telescope are recorded. This has occurred both in the visible spectrum and the near-infrared. These huge detector arrays, combined with wide-field telescopes, provide a veritable deluge of observations, taking us into a new era of astronomy.

The Sloan Digital Sky Survey (SDSS, see Chapter 3) was the first major survey to take advantage of the possibilities of massive detector arrays. Operating in the visible and near infrared, the survey telescope used a detector array with 120 megapixels, cooled by liquid nitrogen. In 2014 the survey was producing 200 gigabytes of data per night of observation, and the catalogued data included more than 500 million photometric measurements and 1 million spectroscopic measurements of individual stars. The 2.5-metre SDSS telescope is located at Apache Point Observatory at latitude 34° N. In the southern hemisphere the nearest equivalent is the 4-metre Visible and Infrared Survey Telescope for Astronomy (VISTA), located at the ESO observatory on Parana at 25° S, and operating since 2009. VISTA is a larger telescope than SDSS, and is designed to observe faint objects including galaxies at large distances. The larger diameter brings greater sensitivity, but inevitably brings also a smaller field of view. The camera therefore needs half the number of pixels than the SDSS, each pixel covering 0.33 arcsecond. A five-year programme of surveys is envisaged, including a survey of a complete hemisphere of the southern sky, and repeated scans of selected areas such as the Magellanic Clouds. The next in line of these survey telescopes will be the 8-metre Large Synoptic Survey Telescope (LSST, Chapter 3), which combines a large field of view with great sensitivity, producing an even larger data rate.

The record number of detector elements is installed in Gaia, the astrometric spacecraft, with nearly 1 billion pixels. The main purpose of Gaia is to observe 1 percent of the 100 billion stars in the Milky Way galaxy, finding their positions and velocities so as to construct a three-dimensional map of the whole galaxy. The survey will be repeated seventy times, gaining sensitivity and measuring transverse velocities. The total flow of data, including photometry and spectroscopy, is on a huge scale, comparable with the data flow from the large terrestrial nuclear research machines such as the Large Hadron Collider. Over its lifetime the spacecraft will transmit to Earth more than 200 petabytes (2×10^{20}) of data, which will be stored, compressed, analysed, and referred to by generations of astronomers.

Multibeam Synthesis

The evolution of optical and infrared telescopes to massive multipixel imaging is matched at radio wavelengths by aperture synthesis. The extended arrays of radio telescopes in the Very Large Array (VLA) and the Multi Element Radio Linked Interferometer Network (MERLIN) (see Chapter 9), and the Giant Metrewave Radio Telescope (GMRT), produce maps with many picture elements, analogous to the multipixel output of the optical-survey telescopes. These radio telescopes are now operating with fibre-optic connections between the telescope elements, providing greater bandwidths and demanding faster operation and increased throughput from the correlator computers. Their success has led to the development of several more synthesis arrays, extending the wavelength coverage to both ends of the available radio spectrum. Very Long Baseline Interferometry (VLBI) has extended to world-wide baselines, producing the greatest angular resolution of any spectral regime. Even the Atacama Large Millimeter Array (ALMA) will be used as part of a global millimetre Very Long Baseline Array (VLBA), able to map very bright objects with a precision of a few tens of microarcseconds. This corresponds to a few times the Schwarzchild radius of the black holes at the centre of the Milky Way or the nearby active galaxy Virgo A (M87), and will be used to test theories of General Relativity.

At the shortest millimetre wavelengths, ALMA synthesizes a telescope aperture extending to tens of kilometres, producing maps with an angular resolution of a 0.01 arcsecond, while observing a patch of sky 10 arcseconds across. This is equivalent to producing an image with 1 million pixels. As this is done with ALMA by bringing together signals from sixty component telescopes, it involves massive computation, but in practice the capabilities of such complex synthesis telescopes is often limited by the available computer power. At the long-wavelength end of the radio spectrum, the Low Frequency Array (LOFAR) is even more dependent on computer power. There are 20,000 individual small antennas, at forty-eight stations spread over much of Europe and linked by fibre-optic cables. The antennas themselves do not move; the telescope beam is formed entirely within the electronic correlator (LOFAR has been called the Software Telescope). The computer installed in 2005 was the second largest in Europe, and by 2014 was already updated, providing even higher computer power.

Projects and Prospects

Massive computer power is at the heart of all major telescope projects now under construction or as designs and aspirations. Even the expansion of single aperture optical telescopes from the 4-metre class to 8 metres and now up to 39 metres, which might seem only an extension of mechanical engineering and mirror construction, would be useless without active optics on an unprecedented scale. Images with diameters a hundred times smaller than uncorrected atmospheric seeing are possible with this large aperture, but they are achievable only if the wavefront converging on the image is divided into thousands of individually controlled segments, each corrected by computer every few milliseconds. With active optics, the images produced by the Thirty Metre Telescope (TMT) and the 39-metre European Extremely Large Telescope (E-ELT) will be only a few milliarcseconds across.

The evolution of the optical telescope over four centuries is shown in Figure 64. The steady progress in aperture, shown on a logarithmic scale, is approximately a doubling in size every thirty years. The light-collecting

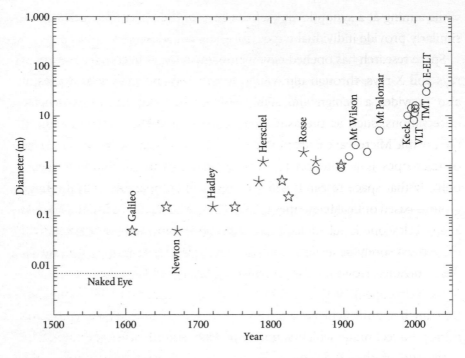

Fig. 64. The progress of telescope apertures, from Galileo to the Thirty Metre Telescope (TMT) and the 39-metre European Extremely Large Telescope (E-ELT). Refracting telescopes are shown by star symbols, speculum reflectors by asterisks, and glass reflectors by open circles.

area has increased by a factor of more than 500,000, and with the dramatic increase in efficiency of detectors, particularly over the last century, the overall performance of telescopes has improved out of all recognition. At the same time, there has been a divergence between the aims of the largest telescopes and the design of the survey telescopes, since large diameter and large field of view cannot easily be combined in a single instrument. The field of view of the E-ELT will be only 10 arcminutes, though to cover that area of sky fully will require a detector array with up to 1 billion pixels. The aim is to produce images of the faintest and most distant individual objects. In contrast, the 4.1-metre VISTA has a field of view of 1.65°, and surveys a large area of sky, identifying objects for the E-ELT and TMT to study at the highest angular resolution and greatest sensitivity. The astrometric Gaia, while aiming at a synoptic view of the stars of the Milky Way galaxy, will similarly provide individual targets for the giant telescopes.

Space research has opened new windows on the Universe, from gamma rays and X-rays, through ultraviolet, to infrared and millimetre waves. It also provides a benign and stable environment for the most delicate observations, such as the astrometry of Gaia and the measurements of the Cosmic Microwave Background by WMAP and Planck. However, if size of telescopes is a criterion for further advances in capability, it seems unlikely that space research will see a growth comparable with the huge ground-based optical telescopes E-ELT, TMT, VISTA, and LSST. The Hubble Space Telescope is scheduled to continue operating only until 2020, and there must surely be a successor with a similar observational capability in the ultraviolet. However the present prospect is only for the James Webb Space Telescope (JWST), operating in the near-infrared. This telescope, due for launch in 2018, achieves a 6.5-metre aperture by unfolding a segmented mirror—a technique which is not easily extended to larger apertures. The advantage of the JWST over the new ground-based telescopes is the extension of infrared wavelength coverage beyond the near-infrared to the longer wavelengths which do not penetrate the atmosphere. This is a rich field, including comparatively cold objects such as planets, and gas clouds from which galaxies are formed, and the exoplanets that have recently been shown by the space telescope Kepler to be a common feature of Milky Way stars.

In 2014 the European Space Agency selected for launch in 2024 a new spacecraft mission named Planetary Transits and Oscillations (PLATO), which would build on the success of Kepler. This spacecraft will have thirty-four cameras, each operating like Kepler's single camera, allowing a total of 1 million stars to be monitored. Again, with a camera having a total number of 700 megapixels the basic rate of data is so large that on-board processing will be essential.

Astronomy at long-wavelength infrared is completely dependent on spacecraft, needing a very cold environment and freedom from atmospheric absorption and thermal emission. The mirror of the JWST will be cooled to 50K by shading from the Sun at its distant L2 location, and will observe efficiently from the near-infrared to the mid-infrared. Extending to the far-infrared is possible with an even colder mirror. This is planned for the Japanese spacecraft Space Infrared Telescope for Cosmology and Astrophysics (SPICA), which will use a 6,000-element array of the very sensitive transition edge detectors (Chapter 5) which must be cooled to below 1K. The 3.5-metre silicon carbide mirror will also be actively cooled.

The high-energy fields of X-rays and gamma rays also depend completely on space telescopes. The focusing technique of the nested Wolter reflectors has given us the three currently functioning X-ray telescopes XMM-Newton, Chandra, and NuSTAR, which provide imaging in the low-energy X-ray band. Extending to higher energies can be achieved only by using geometrical devices such as collimators or the coded mask technique, as demonstrated with Swift, whose burst detector operates at energies up to 150 keV. ESA has chosen to develop an X-ray spacecraft observing at comparatively low energies around 1 keV: the Advanced Telescope for High Energy Astrophysics (ATHENA), which will have a large aperture of 2 square metres, and overall sensitivity improved by an order of magnitude (Chapter 6).

Another proposal to ESA, for a Large Observatory for X-ray Timing (LOFT), would be for higher energies, using a coded mask technique. This proposal is unlikely to be developed until the late 2020s.

The Square Kilometre Array

In a similar way to the transformation of optical and infrared observational astronomy by the multi-element detector arrays, radio (including millimetre-wave) astronomy has been transformed by the expansion of multi-element telescope arrays such as VLA, LOFAR (Chapter 9) and ALMA (Chapter 10). These synthesis arrays have only become possible by the advent of fast digital electronics and massive computer power. The outcome for telescopes both in optical and radio domains is the simultaneous use of multiple beams.

In the optical case these are derived from a single large aperture with a closely packed mosaic of pixels. In contrast, the synthesis radio telescopes have multiple apertures, and the beams are formed by combining the signals electronically. The extreme example of this synthesis technique is the Square Kilometre Array (SKA)—a huge radio telescope which is now being developed, and parts of which are already being brought into use as SKA Pathfinders.

The name of the SKA refers to the total collecting area of the many thousands of telescope elements, which will be spread over a much larger area. The collecting area will be more than a hundred times greater than that of the GBT, which is the largest steerable dish likely to be built in the foreseeable future. Not only that, but the SKA will operate with hundreds of beams simultaneously, and on multiple frequencies. The number of simultaneous observations will be limited only by the processing capacity of the central computers, whose exacting specification is based on developments expected during the next few years.

The SKA is being built in two locations, in Western Australia and South Africa. These will cover two separate frequency bands: 30–350 MHz (SKA LOW, wavelength approximately 10 metres to 1 metre) and 350 MHz to 14 GHz (SKA MID, approximately 1 metre to 2 cm). In both locations there are already radio telescope projects which will establish techniques to be incorporated into the full-sized arrays. At the Murchison site in Western Australia there are two such projects: the Murchison Widefield Array (MWA) and the Australian SKA Pathfinder (ASKAP). The MWA is a long-wavelength array somewhat similar to LOFAR, but its 8,000 basic dipole elements are in a more compact formation, with a core concentration within 1.5 km and extensions out to 3 km. The long-wavelength SKA LOW array will eventually comprise about 1 million individual elements, using a novel design to cover the wide wavelength range (Figure 65). The dishes of the SKA MID antennas will use a new technique which provides each with multiple beams. This is the phased array feed, which replaces the usual single feed at the focus. This technique is being developed initially in Australia, in ASKAP—an array of thirty-six dishes. Construction of the

Fig. 65. Elements of the Square Kilometre Array, SKA LOW, showing a concentration of the dipole elements, of which there will be some hundreds of thousands.

full-scale SKA MID, in South Africa, has started with an array of sixty-four dishes, each 13.5 metres in diameter. This array, engagingly named Meer-KAT, will eventually be incorporated into the SKA, which will have several thousand such dishes, most of which will be located in South Africa, with some distributed over eight African partners.

The scale of the full SKA raises formidable engineering problems. Although about half of the telescope elements will be concentrated in a central area about 5 km across, the other half will be spread at distances up to 3,000 km and be located in many different countries. Connecting the dishes and dipole arrays, with their separate receivers, to the central computer system will require a network of fibre-optic cables with a total length sufficient to stretch round the Earth several times. The rate of raw data at SKA LOW alone will be 157 terabytes/sec—five times the estimated global Internet traffic in 2015—but this must be reduced progressively as the signals are combined and fed to the central computer.

The SKA is organized as an independent international consortium, with the following eleven members: Australia (Department of Innovation, Industry, Science and Research); Canada (National Research Council); China (National Astronomical Observatories of the Chinese Academy of Sciences); Germany (Federal Ministry of Education and Research) (withdrew in 2014); India (National Centre for Radio Astrophysics); Italy (National Institute for Astrophysics); New Zealand (Ministry of Economic Development); South Africa (National Research Foundation); Sweden (Onsala Space Observatory); The Netherlands (Netherlands Organization for Scientific Research); United Kingdom (Science and Technology Facilities Council).

Many other organizations are also involved in the development of the project. The headquarters office is at Jodrell Bank Observatory of the University of Manchester, in Cheshire, England.

Astronomy Transformed

In the mid-twentieth century, when I started research in astronomy, observing was a very personal affair. In my subject of radio astronomy we built our own apparatus, made our observations, and wrote papers and a thesis, nominally within a space of three years. Optical astronomy involved working all night—often travelling to observatories with one's own apparatus and bringing back one's own photographs or a few pages of data. This was a romantic era: Sir Richard Woolley, a former Astronomer Royal, used to say that no-one could call himself an astronomer who had not seen the light of dawn after a hard night's observing. He was also very wary of electronics in his telescope domes, since it distanced an observer from his direct contact with the cosmos. Observers used to ride in a capsule at the prime focus of telescopes such as the 200-inch on Mount Palomar, taking photographs. The contrast today is dramatic. Now it is rare for an astronomer to be present at a telescope while it is taking data for his or her research. Instead, observing is remote or by proxy, and research often involves only accessing the wealth of data already available in the databanks which pour out from the new telescopes whose development is traced in this book.

The transformation started with space research. Manned space laboratories are of little use in astronomy, especially where there is a requirement for continuous undisturbed operation. The small pioneering spacecraft which opened the spectrum to the infrared and ultraviolet were initially the work of small groups of individuals who could see their projects through from an original idea to the assembly of apparatus, working with flight engineers, analysing results, and writing papers, all within a period of a few years. Space projects now take tens of years, involve hundreds of people, and cost hundreds of millons of dollars. But they do produce a vast quantity of data.

In ground-based astronomy the transformation began with the survey telescopes, such as the Schmidt sky surveys, which acquired and stored large quantities of data on their photographic plates, many of which were subsequently scanned by measuring machines and reduced to digitized databanks. With the advent of billion-pixel detector arrays, and with detector efficiencies so much improved that almost every photon entering the telescope is recorded, the volume and quality of data from ground-based astronomical telescopes has increased and improved by orders of magnitude. For many astronomers, research has become a sedentary occupation, accessing databanks rather than looking at the sky. The whole organization of astronomy has changed over the past half century, reflecting a transformation in scale and importance both in academic and popular culture.

In the early twentieth century, every major country had a national observatory devoted to timekeeping and navigation. Many universities had their own observatories, usually organized and staffed separately from their physics research laboratories. There was little enthusiasm in these established observatories for exploring new fields, and the initiatives came instead mainly from university physics departments. Research funding followed a similar path: for example, in the UK the Royal Greenwich Observatory, which had been funded by the Royal Navy because of the requirement for accurate navigation, was transferred in 1965 to a newly formed Science Research Council. The demands of the new astronomies could then be funded on a national scale, with university departments providing most of the initiatives. The relationship between national institutions and universities developed in various ways: in Australia, for example, radio astronomy was, and still is, a continuing activity of the Radiophysics Laboratory, which was a centre for radar research in wartime, while separate centres at Sydney University and the optical observatory at Canberra all competed for research staff and funding. In the USA, a rapidly growing interest in astronomy led the National Science Foundation to bring together universities to create common facilities for research, with the prospect of building much larger telescopes than they could hope for individually. The resulting Association of Universities for Research in Astronomy (AURA) initially had seven member universities, and today comprises thirty-nine institutions. In 1973, AURA opened a 4-metre telescope at the Kitt Peak National Observatory, and the design of this telescope was copied for the 3.9-metre Anglo-Australian Telescope, completed in 1974.

No other country could bring together on its own such a consortium of institutions to create a first-class observatory. International collaboration, however, comes easily to astronomers, and a European grouping was a natural step. The inspiration for the association of astronomers from European countries into the European Southern Observatory (ESO) came from the Dutch astronomer Jan Oort (1900–1992). In 1953 he was already discussing with Walter Baade (1893–1960) the possibilities for countries in the northern hemisphere of gaining access to the southern sky by joining together to develop a new observatory on an excellent site. In 1962, five

European countries formed ESO, which enabled every member to join in front-line research, and today there are fifteen members. The UK was a late joiner, as it already had its own international plans. In 1962, the UK was preparing to observe the southern skies in partnership with Australia, and from 1974 it was involved in a collaboration with Spain for the development of a new northern observatory on the island of La Palma in the Canary Islands. The UK joined ESO in 2002.

Radio astronomy followed a different pattern of international collaboration. Until new technologies opened up the shorter millimetre wavelengths, requiring high dry sites to avoid atmospheric water vapour, there was no need for radio astronomers to build their telescopes outside their own countries; in fact some, like Jodrell Bank Observatory, are still operating not far from their parent universities. However, the possibility of linking radio telescopes in interferometric networks some hundreds or thousands of kilometres apart has fostered a remarkable era of international collaboration. Ten or more observatories may combine to offer observing time on their telescopes for joint projects which may be proposed by any individual and whose scientific merit is judged jointly by the members of the consortium. Observing time is donated by each observatory, with no exchange of funds and no requirement for any return apart from a simple acknowledgement in the published results.

Another world-wide collaboration of radio observatories is using a network of large radio telescopes to measure the arrival times of pulses from an array of pulsars distributed over the sky. The objective is to detect cosmic gravitational waves, which are expected to delay or advance pulse arrival times by amounts of around 10 nanoseconds, varying with periodicities of order several months or years. This demanding level of precision can be achieved only by a geographic spread of observatories using carefully coordinated techniques. Three groups of radio telescopes, in Australia, Europe, and North America, are engaged in this search, and their observations are combined in the International Pulsar Timing Array (IPTA).

Sharing observing facilities has the enormous advantage that astronomical research has become available to every university and institution throughout the world, recruiting talent and expertise whose effect on the development of astronomy has been dramatic. The benefits are seen today in the new techniques translated from fundamental physics research into new instrumentation for all wavebands of the electromagnetic spectrum. We may look back with nostalgic regret at the days (and nights) when research in astronomy involved gazing at the sky through a telescope of one's own construction; that is now the privilege and pleasure of the amateur. Today, the whole cosmos is available to us through the Internet, with databases covering the whole spectrum from radio to gamma. (But yet not the whole of the cosmos; remember the 95 percent in the form of Dark Matter and Dark Energy!) When the surveys have shown us where to look, anyone in the world can join in using the magnificent new telescopes which are our new eyes on the sky.

NOTES

Chapter 1

1. See Henry C. King, *The History of the Telescope*. Charles Griffin and Co., 1955; reprint, Dover, 1979.
2. I recommend Albert van Helden's translation of *Siderius Nuncius*. Chicago: University of Chicago Press, 1989.

Chapter 2

1. Derek Howse, *Greenwich Observatory: Vol 3. The Buildings and Instruments*. Taylor and Francis, 1975.
2. Sir George Biddell Airy, Astronomer Royal from 1835 to 1881.
3. The construction of the Palomar 200-inch is well described in Ronald Florence, *The Perfect Machine*. Harper Perennial, 1995.
4. The UK withdrew from the original collaboration in 2010, after which, the AAO was renamed the Australian Astronomical Observatory.

Chapter 3

1. Jerry Nelson moved to the University of California at Santa Cruz in 1994, and there, in 1999, he created the Center for Adaptive Optics.
2. A.-M. Lagrange *et al.*, 'A giant planet imaged in the disk of the young star β Pictoris', *Science*, 329 (2010): 57.
3. S. Gillessen *et al.*, 'Monitoring stellar orbits around the massive black hole in the galactic center', *Astrophysical Journal*, 692 (2009): 1075.
4. The design of the LSST is based on a proposal by Roderick Willstrop in 1984. R. V. Willstrop, 'The Mersenne-Schmidt: A three-mirror survey telescope', *Monthly Notices of the Royal Astronomical Society*, 210 (1984): 597.

Chapter 4

1. 2MASS is a joint project of the University of Massachusetts and the Infrared Processing and Analysis Center/California Institute of Technology, funded by the National Aeronautics and Space Administration and the National Science Foundation.

Chapter 5

1. Robin H. Brand, *Britain's First Space Rocket: The Story of the Skylark*. YPD Books, 2014.
2. See 'Sir Robert Wilson, CBE', *Biographical Memoirs of Fellows of the Royal Society*, 50: 367–86.
3. A byte of information usually comprises about eight digits, or bits. A data rate of 5 megabits per second is a typical rate for a domestic connection to the Internet.

Chapter 6

1. J. F. Smith and G. M. Courtier, 'The Ariel V programme', *Proceedings of the Royal Society A*, 350 (1976): 421–39.
2. M. Bachetti *et al.*, 'An ultraluminous X-ray source powered by an accreting neutron star', *Nature*, 514 (2014): 202.
3. S. P. Tendulkar *et al.*, 'NuSTAR discovery of a cyclotron line in the Be/X-ray binary RX J0520.5-6932 during outburst', *Astrophysical Journal*, 795 (2014): 154.

Chapter 7

1. J. van Paradijs, 'Transient optical emission', *Nature*, 386 (1997): 686.
2. M. R. Metzger, 'Redshift of the optical lines', *Nature*, 387 (1997): 878.
3. N. R. Tanvir *et al.*, 'A glimpse of the end of the dark ages: the gamma-ray burst of 23 April 2009 at redshift 8.3', *Nature*, 461 (2009): 1254.
4. R. Browning, D. Ramsden, and P. L. Wright, 'Detection of pulsed gamma radiation from the Crab Nebula', *Nature Physical Science*, 232 (1971): 99.
5. Gamma rays from the Crab pulsar were discovered during a balloon flight; see B. McBreen *et al.*, 'Pulsed high-energy gamma rays from the Crab Nebula', *Astrophysical Journal*, 184 (1973): 571–80. Gamma rays from the Vela pulsar were discovered by the spacecraft SAS-2; see D. J. Thompson *et al.*, 'SAS-2 high-energy gamma-ray observations of the Vela pulsar', *Astrophysical Journal Letters*, 200 (1975): 79–82.

Chapter 8

1. Jansky's antenna was designed by Edwin Bruce at the Bell Telephone Laboratories. The pattern is known as a Bruce array.
2. Beamwidth is usually defined as the angular distance between points at which the sensitivity is half that at the peak.
3. For references and details of these pulsar discoveries, see Andrew Lyne and Francis Graham-Smith, *Pulsar Astronomy*, 4th edn. Cambridge University Press, 2012.
4. Details of FAST are available in a paper by R. Nan *et al.*, 'The Five-Hundred-Meter Aperture Spherical Radio Telescope (FAST) project', *International Journal of Modern Physics D*, 20 (2011): 989.

Chapter 9

1. The term 'fringes' derives from another diffraction phenomenon: the fringes of light seen at the edges of a shadow.
2. Martin Ryle was Astronomer Royal from 1972 to 1982.
3. F. G. Smith, 'The measurement of the angular diameter of radio stars', *Proceedings of the Physical Society B*, 65 (1952): 971–80.
4. F. G. Smith, 'Apparent angular sizes of discrete radio sources: Observations at Cambridge', *Nature*, 170 (1952): 1065.
5. J. H. Blythe, 'A new type of pencil beam aerial for radio astronomy', *Monthly Notices of the Royal Astronomical Society*, 117 (1957): 644–51.
6. This simple form of aperture synthesis works only for sites away from the equator.
7. Martin Ryle and Ann C. Neville, 'A radio survey of the north polar region with a 4.5 minute of arc pencil-beam system', *Monthly Notices of the Royal Astronomical Society*, 125 (1962): 39–56.
8. L. L. McCready, J. L. Pawsey, and R. Payne-Scott, 'Solar radiation at radio frequencies and its relation to sunspots', *Proceedings of the Royal Society A*, 190 (1947): 357–75.
9. I. Rosenberg, 'A high-resolution map of Cassiopeia A at 2.7 GHz', *Monthly Notices of the Royal Astronomical Society*, 147 (1970): 215–30.
10. N. W. Broten et al., 'Observations of quasars using interferometer baselines up to 3,074 km,' *Nature*, 215 (1967): 38. B. G. Clark, 'A review of the history of VLBI', in J. Zensus et al., *Radio Astronomy at the Fringe*, Astronomical Society of the Pacific Conference Series, 300 (2003): 1–8.
11. M. L. Lister et al., 'The Pearson–Readhead survey of compact extragalactic radio sources from space. I. The images', *Astrophysical Journal*, 554 (2001): 948.

Chapter 10

1. A. E. E. Rogers et al., 'Observations of the 327 MHz deuterium hyperfine transition', *The Astronomical Journal*, 133 (2007): 1625.
2. For an account of the quantum physics of hydroxyl and more complex molecules, see B. F. Burke and F. Graham-Smith, *An Introduction to Radio Astronomy*, 3rd edn. Cambridge University Press, 2010.
3. The observations of maser radiation from NGC 4258 are described by J. R. Herrnstein et al., 'A geometric distance to the galaxy NGC 4258 from orbital motions in a nuclear gas disk', *Nature*, 400 (1999): 539–41.
4. R. Maiolino et al., 'The assembly of "normal" galaxies at $z \approx 7$ probed by ALMA', *Monthly Notices of the Royal Astronomical Society*, 452 (2015): 54–68.
5. E. L. N. Jensen and R. Akeson, 'Misaligned protoplanetary disks in a young binary star system', *Nature*, 511 (2014): 567–9.
6. A. I. Sargent and S. Beckwith, 'Kinematics of the circumstellar gas of HL Tauri and R Monocerotis', *Astrophysical Journal* 323 (1987): 294–305.

7. The first observation of a gravitational lens: see D. Walsh, R. F. Carswell, and R. J. Weymann, '0957+561 A, B: twin quasistellar objects or gravitational lens?', *Nature*, 279 (1979): 381–4.

8. D. A. Riechers *et al.*, 'A dust-obscured massive maximum-starburst galaxy at a redshift of 6.34', *Nature*, 496 (2013): 329–33.

Chapter 11

1. V. M. Slipher, 'Nebulae', *Proceedings of the American Philosophical Society*, 56 (1917): 403–9.

2. The Hubble Constant is not, strictly speaking, a constant. The linear relation between redshift and distance is not followed at large cosmological distances.

3. G. Lemaître, 'A homogeneous Universe of constant mass and increasing radius, accounting for the radial velocity of extragalactic nebulae', *Annales de la Société Scientifique de Bruxelles*, A47 (1927): 29–39.

4. R. A. Alpher and R. C. Herman, 'Evolution of the Universe', *Nature*, 162 (1948): 774–5.

5. In the Michelson spectral interferometer, the light transmitted by the pair of mirrors depends on the number of wavelengths between them. The effect is to multiply the spectrum by a sine wave with a periodicity depending on the spacing between the mirrors. The spectrum is obtained from a Fourier transform of the amplitude of the sine wave as measured over the whole range of spacings.

6. P. de Bernardis *et al.*, 'A flat Universe from high-resolution maps of the cosmic microwave background radiation', *Nature*, 404 (2000): 955–9.

7. The Wilkinson Microwave Anisotropy Probe (WMAP) was originally MAP. The W denotes David Wilkinson (1935–2002), one of the chief designers.

8. For a technical description of WMAP, see B. F. Burke and F. Graham-Smith, *An Introduction to Radio Astronomy*, 3rd edn. Cambridge University Press, 2010: ch. 15.

9. See P. A. R. Ade *et al.*, 'A measurement of the cosmic microwave background B-mode polarization power spectrum at sub-degree scales with POLARBEAR', *Astrophysical Journal*, 794 (2014): 171.

10. The 2011 Nobel Prize was award to Saul Permutter, Brian Schmidt, and Adam Reiss.

11. The array in South Africa is the Precision Array for Probing the Epoch of Reionization (PAPER); see A. R. Parsons *et al.*, 'The Precision Array for Probing the Epoch of Reionization: 8 station results', *Astronomical Journal*, 139 (2010): 1468. In Australia the array is the Murchison Widefield Array (MWA); see S. J. Tingay, 'The Murchison Widefield Array: The Square Kilometre Array precursor at low radio frequencies', *Publications of the Astronomical Society of Australia*, 30 (2013): 7.

FURTHER READING

Coles, Peter. *Cosmology: A Very Short Introduction*. Oxford University Press, 2001. Covers both the history of cosmology and the latest developments up to the year 2000, using simple language without mathematics.

Giacconi, Riccardo. *Secrets of the Hoary Deep*. Johns Hopkins University Press, 2008. A personal history of modern astronomy, by the pioneer of X-ray astronomy, describing the development of X-ray telescopes from 1959 to 2006. Giacconi also provides a detailed account of the Hubble Space Telescope

Graham-Smith, Francis. *Unseen Cosmos*. Oxford University Press, 2013. An introduction to radio astronomy and its history up to 2012.

Hearnshaw, John. *The Analysis of Starlight*. Cambridge University Press, 2014. The story of the analysis of starlight by astronomical spectroscope, from Joseph Fraunhofer's discovery of spectral lines in the early nineteenth century through to the year 2000. Both observational and theoretical aspects are described in a non-mathematical framework.

Hoskin, Michael. *The History of Astronomy*. Oxford University Press, 2003. Focuses on the major developments of the seventeenth, eighteenth, and nineteenth centuries: Copernicus, Kepler, Galileo, Newton, and beyond.

K. I. Kellermann and J. M. Moran, 'The development of high-resolution imaging in radio astronomy', *Annual Review of Astronomy and Astrophysics*, 39 (2001): 457–509.

King, Henry. *The History of the Telescope*. Charles Griffin and Co., 1955; reprint, Dover, 1979. The classic work on telescopes up to the time of the Palomar 200-inch reflector.

Longair, Malcolm. *The Cosmic Century*. Cambridge University Press, 2006. A comprehensive history of astrophysics and cosmology in the twentieth century, including radio, X-rays, and gamma rays.

Morison, Ian. *An Amateur's Guide to Observing and Imaging the Heavens*. Cambridge University Press, 2014.

Rowan-Robinson, Michael. *Night Vision: Exploring the Infrared Universe*. Cambridge University Press, 2013. A survey of progress in infrared observation, and an introduction to the pioneering scientists and engineers who have painstakingly developed infrared astronomy over two hundred years.

INDEX